# Engineering Materials

This series provides topical information on innovative, structural and functional materials and composites with applications in optical, electrical, mechanical, civil, aeronautical, medical, bio- and nano-engineering. The individual volumes are complete, comprehensive monographs covering the structure, properties, manufacturing process and applications of these materials. This multidisciplinary series is devoted to professionals, students and all those interested in the latest developments in the Materials Science field, that look for a carefully selected collection of high quality review articles on their respective field of expertise.

**Indexed at Compendex (2021) and Scopus (2022)**

Raden Maizatul Aimi Mohd Azam ·
Hamidah Mohd Saman

# Self-Healing Concrete

## Encapsulated *Geobacillus stearothermophilus* and Biomineralised Alginate Hydrogel Systems

Springer

Raden Maizatul Aimi Mohd Azam
Faculty of Civil Engineering
Universiti Teknologi MARA (UiTM)
Shah Alam, Selangor, Malaysia

Hamidah Mohd Saman
Faculty of Civil Engineering
Universiti Teknologi MARA (UiTM)
Shah Alam, Selangor, Malaysia

ISSN 1612-1317 ISSN 1868-1212 (electronic)
Engineering Materials
ISBN 978-981-95-7552-7 ISBN 978-981-95-7553-4 (eBook)
https://doi.org/10.1007/978-981-95-7553-4



This Springer imprint is published by the registered company Springer Nature Singapore Pte Ltd.
The registered company address is: 152 Beach Road, #21-01/04 Gateway East, Singapore 189721, Singapore

# Preface

As an academic in concrete technology, we were inspired to write this book to share our expertise on self-healing concrete with researchers, professionals, students, and practitioners. Self-healing concrete presents significant advantages, particularly in reducing environmental impact and advancing sustainable practices within the construction industry. Through this comprehensive guide, we aim to contribute to the development of innovative materials and promote sustainability and environmental stewardship in concrete technology.

This book is organised into five chapters, each exploring various aspects of self-healing concrete. It focuses on different self-healing mechanisms and types of healing agents, with a particular emphasis on microbiologically induced calcite precipitation (MICP). The book also investigates the performance of self-healed concrete using a range of analytical tools, highlighting the role of alginate hydrogel-encapsulated *Geobacillus stearothermophilus* as an effective self-healing agent.

Chapter 1 introduces the concept of bio-based self-healing in concrete and its relevance to sustainable development in the construction industry.

Chapter 2 explains the mechanisms of MICP-based self-healing, detailing the biochemical processes involved in biomineralisation. It discusses various bacterial species employed in self-healing studies and examines key factors influencing biomineralisation, particularly those related to crack filling.

Chapter 3 explores three application methods of bacterial self-healing agents in concrete: direct incorporation, surface treatment, and encapsulation prior to integration into concrete or mortar. The chapter evaluates the effects of these approaches on the physical, mechanical, and durability properties of concrete, drawing from both existing literature and the authors' experimental findings.

Chapter 4 reviews techniques used to assess self-healing efficiency, focusing on MICP-based systems. These methods are grouped into three main categories: visual inspection, morphological analysis, and confirmation of calcite precipitation. Key analytical tools include Scanning Electron Microscopy (SEM), Thermogravimetric Analysis (TGA), and X-Ray Diffraction (XRD).

Chapter 5 examines the physical and functional characteristics of encapsulated *Geobacillus stearothermophilus*, evaluating its viability and performance as a self-healing agent. The chapter demonstrates the bacterium's ability to survive within alginate hydrogel and precipitate calcium carbonate ($CaCO_3$) when embedded in cementitious materials. These findings form a foundation for understanding microbial self-healing mechanisms and contribute valuable insights to the body of knowledge for future research.

Shah Alam, Malaysia

Raden Maizatul Aimi Mohd Azam
Hamidah Mohd Saman

**Acknowledgements** First and foremost, I would like to express my sincere thanks to Associate Prof. Dr. Rohana Hassan, Dr. Arni Munira Markom, and Prof. Dr. Aidah Jumahat for their invaluable guidance and encouragement in completing this book. I would also like to extend my heartfelt gratitude to my mentors, Prof. Dr. Azmi Ibrahim, Prof. Ir. Dr. Suhaimi Abd Talib and Prof. Ir. Dr. Che Khairil Izam Che Ibrahim for their insightful guidance, continuous motivation, and thoughtful advice, which have greatly enriched my knowledge and shaped the direction of this work.

I would like to acknowledge my ex Ph.D. student, and also my colleague Ir. Ts. Dr. Raden Maizatul Aimi Mohd Azam for the materials contributed in this book. This book will not be completed without her contribution. I also would like to express my gratitude to Universiti Teknologi MARA (UiTM) for providing facilities, laboratories and equipment for conducting research in which the review and findings are the major part of this book. Without these facilities, this book would not be possible.

Furthermore, I would like to thanks my colleagues for their love, motivation, encouragement and continuous support. Finally, I would like to thank my family members, my children; Abdullah Mubarak, Fatimah, Safiah and Maimunah. Not forget, my husband Major Mohd Zakuan Mohd Nor for his love, caring, inspiration and continuous support who motivates me when in need. This blessing has enlightened my path towards completing this book.

Lastly, I am very thankful to all researchers whose research papers, thesis, book and articles that have been referred, without these references I may not able to get right path for my book.

**Competing Interests** The authors have no competing interests to declare that are relevant to the content of this manuscript.

# About This Book

This book delves into microbial based autonomous self-healing concrete. It offers an in-depth review that include the mechanisms that derived from microbiologically induced calcite precipitation (MICP) and different application of autonomous bio based healing. This book also covers evaluation of self-healed concrete and properties of encapsulated alginate hydrogel encapsulated *Geobacillus stearothermophilus* as self-healing agent in concrete. This book serves as a valuable reference for both students and researchers in the field. Its unique approach lies in the use of simple, clear English, making it accessible to readers of all levels. By breaking down complex concepts into understandable terms, the book ensures that even those new to the topic can grasp the fundamentals. Additionally, it is highly relevant for practitioners, providing practical insights into different types of bio based self-healing techniques that can be implemented in real-world applications. Whether you are an academic looking to expand your knowledge or a professional seeking practical solutions, this book offers guidance on self-healing concrete, making it an essential resource for anyone interested in this innovative area of construction technology. The authors hope this book will give valuable insights into development of bio-based self healing agents and other potential materials as self-healing agent.

# Contents

# About the Authors

**Raden Maizatul Aimi Mohd Azam** obtained her Diploma in Civil Engineering from Politeknik Sultan Azlan Shah (PSAS) in 2006, followed by her Bachelor of Civil Engineering (Hons.) from Universiti Teknologi MARA (UiTM) in 2009. She later pursued her Master of Science in Civil Engineering (Construction) at UiTM in 2011 and successfully completed her Doctor of Philosophy (Ph.D.) in Civil Engineering at UiTM in 2018.

She began her career as a Graduate Research Assistant at UiTM in 2012 under the Fundamental Research Grant Scheme (FRGS) before serving as a Lecturer at UCSI University in 2017. In 2018, she joined the Public Works Department (JKR) Malaysia as a Civil Engineer under the Civil and Structural Engineering Branch (CKAS), where she was responsible for structural analysis, retrofitting proposals, seismic fragility studies, and rehabilitation of government buildings. In 2023, she returned to UiTM as a Senior Lecturer at the School of Civil Engineering, College of Engineering.

Professionally, Dr. Raden Maizatul Aimi is a registered Professional Technologist (Ts.) with the Malaysia Board of Technologists (MBOT), a Graduate Engineer with the Board of Engineers Malaysia (BEM), and a Graduate Member of the Institution of Engineers Malaysia (IEM). She is also a member of the Concrete Society of Malaysia (CSM) and the American Concrete Institute (ACI) Kuala Lumpur Chapter.

Her main research areas are materials and structural engineering, focusing on self-healing concrete, bio-concrete, and sustainable construction materials. She has contributed to journals, including the *ACI Materials Journal* and *IEM Jurutera Bulletin*, and presented her work at international conferences such as InCIEC and CONCET. She has also participated in technical committees under JKR related to structural design standards, building rehabilitation, and material testing.

Throughout her academic and professional career, she has supervised final-year projects in various type of concrete, fostering research collaboration and mentoring future engineers. She has also been actively involved in technical committees under JKR, including the Committee on Structural Component Standard Design (Eurocode), the Committee of Material Testing Plan for Building Rehabilitation, and the Technical Committee on Composite Building Materials for SIRIM.

**Hamidah Mohd Saman** obtained her Bachelor in Civil Engineering Degree in 1988 from the University of Miami, Coral Gables, United States of America (USA). She worked as a Researcher in SIRIM Berhad from 1992 before furthering her studies at the University of New South Wales, Sydney, Australia, and obtained her Ph.D. in Civil Engineering in 1999. She joined Universiti Teknologi MARA (UiTM) in 2000 as a Lecturer and was promoted to Associate Professor in 2007. She is a Professional Engineer under Board of Engineers Malaysia (BEM). Her active involvement in professional bodies is portrayed by her designation as a Secretary of Concrete Society of Malaysia (CSM) for seven sessions, Committee, Treasurer, Secretary and Board of Director, American Concrete Institute—Malaysia Chapter and was a Council Member of Confederation of Scientific and Technological Associations in Malaysia (COSTAM). Based on her expertise, she has been appointed as a Chief Editor for *Journal of Sustainable Civil Engineering (JSCET)*, *Scientific Research Journal (UiTM)*, a member of the editorial board for *Malaysian Construction Research Journal (CREAM-CIDB)*, *The Institution of Engineers Malaysia (IEM) Journal*, *Journal of Engineering Science and Technology (JESTEC)* and *International Sustainable Construction Journal* co-published by UTHM-Concrete Society of Malaysia (CSM). During her service as an academic member in UiTM, she has published more than 300 papers in journals and conference proceedings at the international and national levels in the area of durability in concrete, lightweight concrete, high-performance concrete and bio-concrete, self-healing concrete, international construction, strengthening and repair materials for concrete structures using non-corrodible materials. Nano, bio-concrete and high fire-resistance cement materials are her new areas of interest. She has been actively involved in research and consultation works and has been granted more than RM 2.0 million over the last 5 years of research grants from government agencies and industries. So far, she has supervised 30 Ph.D. students, and 22 of them have graduated. She was promoted to full Professor in 2015.

# List of Figures

# List of Tables

# Chapter 1
# Self-Healing of Concrete and Its Main Issues

## 1.1 Introduction

Concrete is broadly used as an engineering material in construction because of its excellent mechanical, durability properties and relatively low cost as compared to other building materials [10, 19]. However, concrete is susceptible to progression and coalescence in the micro-crack formation due to its inherent heterogeneity, low tensile strength and non-ideal service environment [7, 30, 32]. This major drawback of concrete does not only lead to low strength concrete with reduction of quality and durability but has become an open pathway for the ingress of water and harmful substances that vulnerable to the concrete. Lots of additives and admixtures are used to ensure the service life of concrete either by improving permeability or filling up cracks.

The variety of treatment techniques is available to repair the concrete cracks with a substantial cost needed. The broad arrays of organic and inorganic products such as epoxy injection [34] are available in the market as treatments to heal cracks in concrete. Nevertheless, the disadvantageous of these remediation products is that it needs constant maintenance with the solvents in the products contributes to pollution and environmental problem [31]. It also requires manual and constant observation to reduce crack propagation as limited availability of access for observation in indiscernible location. For that reason, a self-healing approach has become the authors' interest as a natural way to remediate cracks.

This book presents a self-healing approach using biomineralisation, process derived from Microbially Induced Calcite Precipitation (MICP), as an emerging technology bridging concrete and biology. MICP utilises the ability of urease positive bacteria activity in producing urease enzyme to influence the precipitation calcium carbonate ($CaCO_3$) or calcite. The biomineralisation and the reactions of MICP are further elaborated in Chap. 2. Concrete that adopts bacteria is also known as bio-concrete. It is an innovative method that has a multi-application in engineering [1,



R. M. A. M. Azam and D. H. M. Saman, *Self-Healing Concrete*, Engineering Materials,
https://doi.org/10.1007/978-981-95-7553-4_1

9, 13, 15, 16, 22, 38, 39] and one of them is as crack filling in concrete. The effectiveness of MICP, a crack healing agent in cementitious materials, has been proven in previous works [2, 12, 14, 18, 28, 37]. The calcite ($CaCO_3$) forms oolitic and considerable limestone in sediments [8], making it suitable to be explored as an alternative in crack remediation. This method transcends other repair technique as it is bio-based, profitable, durable and eco-friendly [32]. So far, there are three (3) standard methods of applying MICP in concrete crack remediation namely direct incorporation, surface treatment and also encapsulation and these will be elaborated in Chap. 3.

The following sections elaborate further the mechanisms and various concepts of autogeneous and autonomous self healing with the emphasis given to bacteria self healing based. It also highlights key issues related to the use of bacteria as healing agents, prompting further investigation to address these challenges.

## 1.2 Concrete Crack Self-Healing Mechanisms

Cracks are the main reason for a shorter service life of concrete structures. The occurrence of minor cracks would eventually lead to further deterioration degree causing a major break and make it difficult to be repaired. Cracks occurred functions as an entry channel for deleterious, aggressive liquids and gases penetrates into it. Different types of fissures occur at all stages of actual serviceability [47], thus, results in loss of durability and overall quality of concrete structures earlier than expected. Early detection and treatment of crack may substantially reduce deterioration level of embedded steel bar, as the ingression rate of corrosive chemicals such as water and chloride ion into the concrete structures decrease [23, 25]. For that reason, the use of immediate and proper treatment to heal cracks is necessary to avoid extendable of cracks. Self healing concrete refers to the development of innovative technologies and materials that enable concrete to repair its own cracks and damage with or without external intervention. One of the sustainable methods is by making concrete to self-heal by using a suitable healing agent to remediate the occurrence of the crack in indiscernible location. Firstly, the self-healing mechanisms can be broadly categorised into two (2); autogenous and autonomous healing. They are both self-healing mechanisms in concrete, but they differ in their processes as outlined below:

### *1.2.1 Autogenous Healing*

- Definition: This is the natural ability of concrete to heal minor cracks without external intervention.

- Mechanism: It primarily occurs due to the continued hydration of unreacted cement particles and the deposition of calcium carbonate ($CaCO_3$) from the reaction of calcium hydroxide with $CO_2$ and water.
- Limitations: Works best for very small cracks (typically < 0.2 mm) and is dependent on the availability of unhydrated cement and moisture.

### *1.2.2 Autonomous Healing*

- Definition: This involves the deliberate incorporation of healing agents or technologies to repair cracks in concrete actively.
- Mechanism: It can include microcapsules filled with healing agents, vascular networks, shape-memory polymers, or bacteria-based healing using Microbially Induced Calcite Precipitation (MICP).
- Advantages: More effective for larger cracks (> 0.2 mm) and can be engineered for enhanced performance over the concrete's lifespan.

The autogeneous healing relies on the natural hydration of unhydrated cement particles within the concrete to fill cracks. It can be enhanced by incorporating materials like fly ash or silica fume to promote this process. Meanwhile, the autonomous healing include microbial-based healing, chemical additive-based healing and shape memory materials. Micro-based healing uses bacteria (such as *Bacillus*) that produce limestone when exposed to water and nutrients (e.g., calcium lactate). These bacteria are embedded in concrete during mixing. When cracks form and water enters, the bacteria are activated and precipitate calcite, sealing the cracks. Secondly, by including encapsulated healing agents (e.g., epoxy or polyurethane resins) or other reactive chemicals. When cracks occur, the capsules break, and the agents flow into the cracks and harden, effectively sealing them. Thirdly, by using shape memory materials. This approach incorporates materials that "remember" their original shape and can expand or contract to close small cracks. However, it is the interest of authors to explore microbial based healing and become the main subject of this book.

Autonomous healing in concrete also could be achieved through the incorporation of materials such as bacteria, superabsorbent polymers (SAP), or encapsulated healing agents, and it has emerged as a promising advancement in construction technology. These innovative materials enhance the natural self-healing capacity of concrete, enabling it to repair cracks. This concept has captured the interest of the authors, driven by its potential to reduce maintenance costs, extend the lifespan of concrete structures, and support more sustainable and durable construction practices. This book explores the science, mechanisms, and practical applications of autonomous healing in concrete, with particular emphasis on bacteria-based self-healing systems.

## 1.3 Main Issues in Bacteria Self-Healing Based Concrete

Ureolytic bacteria capable of producing urease enzyme are among the bacteria that have been used for self healing agents in concrete to close cracks in concrete. Some other bacterial species such as *Bacillus Megaterium*, *Bacillus licheniformis* and *Bacillus flexus* [32], *Bacillus cohnii* [24, 25], *Sporocarcina pasteurii* [11, 26, 40], *Bacillus sphaericus* [21, 49, 50] and *Bacillus subtilis* [42, 46] directly incorporated into concrete have proven and able to enhance concrete performance. However, the limitation of the bacteria used in such method exhibits temporal activity and a shorter lifetime of bacteria when embedded or mixed up directly to the concrete [27, 40] due to the increase of pH (> 12) in concrete. Therefore, the properties enhancement and healing would not be efficient. Moreover, the decreasing of pore diameter during the hydration [41] of material destroyed the bacterial cell wall [28]. For this reason, the protection of bacteria is vital so that the bacteria and nutrient can act simultaneously with the occurrence of crack [4, 44].

The healing process must begin from inside to outside, in ensuring the concrete could have its self-healing. The selection and the growth condition of bacteria that give significant result in the healing process must include the ability of bacteria to withstand in the harsh environment and capability to precipitate by exploiting available sources in the shortage of nutrition period. The type of bacteria that has been recognised for its survivability and has shown a promising result in previous studies is the kind that can produce endospore such as from genus *Bacillus* and *Clostridium* [5, 28, 35].

Due to high temperature during concrete hydration, the exploration of bacteria dragged into the type that can resist heat and at the same time capable of producing endospore. The *Geobacillus stearothermophilus* is another bacteria from genus *Bacillus* that was selected as a highlight in this book. The *Geobacillus stearothermophilus* (Fig. 1.1) can be found in the geothermally heated region or high-temperature ambient such as hot spring, and it is a highly resistant bacteria due to its capability thriving high temperature (30–75 °C) and [6, 36, 51]. The *Geobacillus stearothermophilus* capable to produce endospore by permitted harsh situation [20, 43]. As stated by Kumar [33] and Zeigler [52], the morphology and resistance properties of *Geobacillus* enable them to be mobilised in the atmosphere and transported within the long distance. On the other hand, the longevity of *Geobacillus stearothermophilus,* enable them to lie quiescent but viable for long periods of time in extreme condition, accumulating gradually over time by achieving surprisingly high population. In the concrete mixing stage and further ongoing hydration which in turn the concrete structure become dense, there might be a risk that the bacterial cells will be damaged and destroyed [30, 48].

Hence, to increase the protection of *Geobacillus stearothermophilus*, encapsulation that will become a bacteria carrier before addition in concrete is preferable. Bacteria can be directly incorporated during mixing or spraying bacteria broth (surface treatment) to heal cracks. Direct incorporation of bacterial (in vegetative cells or endospores) in the concrete mixture during mixing process, however, only

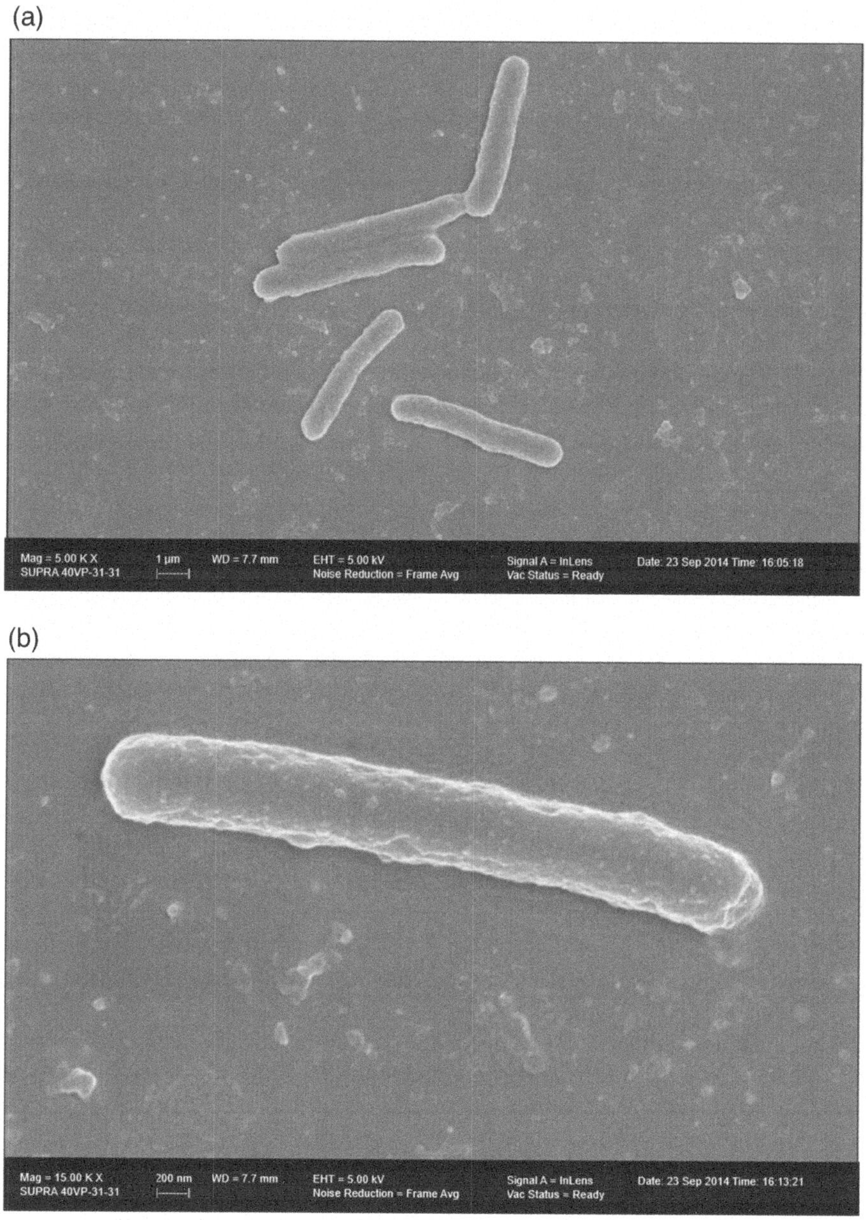


**Fig. 1.1** The morphology of *Geobacillus stearothermophilus* vegetative cells in **a** short chain and **b** in single form

remained viable for a period up to four (4) months [3, 28, 29, 44]. Besides, the mechanical forces during the mixing process and further hydration in concrete microstructure would diminish the lifespan of the spores as pore width decreased below 1 μm, which is the typical size of *Bacillus spores* [30, 48]. Hence, bacteria need to be encapsulated to increase the protective barrier and chances to survive in the harsh concrete environment. The encapsulation will become a bacteria carrier before addition in concrete. The carrier functions as a shell to shield the bacteria from destruction, with no hindering effect on bacterial calcium carbonate precipitation upon cracking and with minimal effect on concrete matrix [48]. The embedment of bacteria in the carrier ensure the bacteria being dormant but viable in the concrete matrix that will aid to self-healing potential [28, 30, 45, 49, 50]. For instance, Wang et al. [49] used the hydrogel encapsulated carbonate as a carrier for *Bacillus* types bacteria in the concrete. Jonkers et al. [28] embedded bacteria and their food in clay capsules while De Belie and De Muynck [17] used sol-gel to increase the protection of bacteria before mixing in the concrete. The encapsulated healing agents will disperse through the mortar matrix. Once cracks appear, the rupture of the hydrogel beads released healing agent into crack thus repair the cracks [45]. Alginate hydrogel is one of many other materials has been chosen to encapsulate the bacteria for self healing purpose. Use of sodium alginate as a vehicle protector for the bacteria loaded with proper nutrient to enable autonomous healing process in the cementitious material has been explored and reported in Chap. 3. Chapter 4 presents the properties and methods to quantify healing efficiency by bacteria; by direct incorporation, surface treament and by encapsulated.

With this background in view, an attempt was also made to observe the effect of incorporating *Geobacillus stearothermophilus*, the *Bacillus* family with sodium alginate (Fig. 1.2) as vehicle protector into mortar matrix. Chapter 5 presents the investigation results on the capability of *Geobacillus stearothermophilus* encapsulated in alginate hydrogel as a self-healing agent in mimiked environment and mortar. The investigation took on two (2) major parts which are; the suitability and survivability of *Geobacillus stearothermophilus* encapsulated in alginate hydrogel and the ability of alginate-encapsulated *Geobacillus stearothermophilus* in mortar crack remediation.

Therefore, the initiative to compare the enhancement and healing properties derived by *Geobacillus stearothermophilus* with other species of bacteria would be beneficial in varying capabilities of bacteria in concrete maintenance.

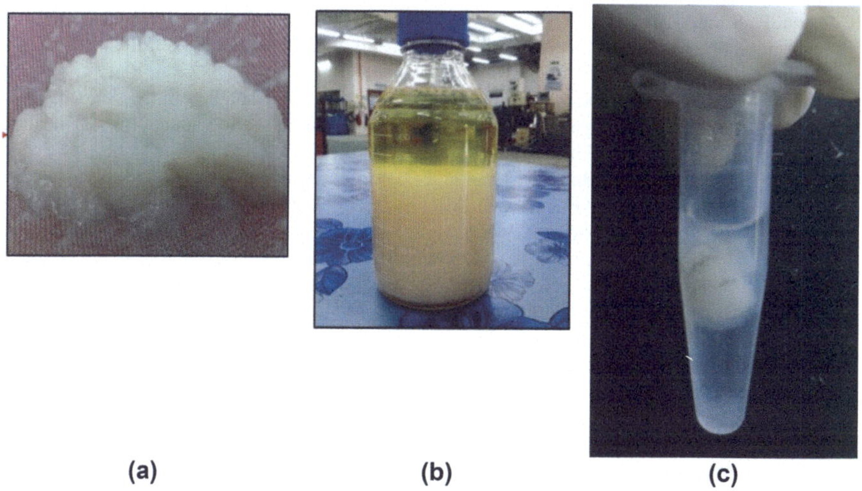


**Fig. 1.2** **a** Rinsed alginate-encapsulated *Geobacillus stearothermophilus*; **b** Alginate-encapsulated *Geobacillus stearothermophilus* immersed in distilled water for use in concrete/mortar mixing and **c** Alginate-encapsulated *Geobacillus stearothermophilus* in phosphate buffer

# References

1. Abo-El-Enein, S.A., Ali, A.H., Talkhan, F.N., Abdel-Gawwad, H.A.: Utilisation of microbial induced calcite precipitation for sand consolidation and mortar crack remediation. HBRC J. **8**(3), 185–192 (2012). https://doi.org/10.1016/j.hbrcj.2013.02.001
2. Achal, V., Siddique, R., Reddy, M., Mukherjee, A.: Improvement in the compressive strength of cement mortar by the use of a microorganism—*Bacillus megaterium*. In: Limbachiya, Kew (eds.) Excellence in Concrete Construction Through Innovation, pp. 27–29 (2009)
3. Ahmad, I., Shokouhian, M., Owolabi, D., Jenkins, M., McLemore, G.L.: Assessment of biogenic healing capability, mechanical properties, and freeze–thaw durability of bacterial-based concrete using bacillus subtilis, *Bacillus sphaericus*, and *Bacillus megaterium*. Buildings **15**(6), 943 (2025)
4. Ahmad, M.A., Zhang, J., Liu, B., Guohao, X., Xiaoyi, T., Haoying, G., Changjie, S., Runhao, L., Xiaona, X., Weilin, L.: Synergistic effect of composite bacteria on self-healing process of concrete crack. Case Stud. Constr. Mater. **20**, e03028 (2024)
5. Ahmed, A.S.S., Elbahnasawy, M.A., Mahgoub, H.A., Maarouf, R.E.: Impact of clostridium sporogenes endospores and boswellic acid on breast cancer induced by N-Nitroso-N-methyl urea (MNU) in female rats. Microb. Biosyst. **10**(1), 152–169 (2025)
6. Aireche, M., Merzoug, M., Hammadi, A. I., Zater, Z. Y., Bendida, K., Brakna, C. N., Berrazeg, M., Aireche, A. Y., Saidi, Y., & Todorov, S. D. (2025). Taxonomic diversity and antimicrobial potential of thermophilic bacteria from two extreme Algerian Hot Springs. Microorganisms 13(6), 1425
7. Akbulut, Z.F., Tawfik, T.A., Smarzewski, P., Guler, S.: Advancing hybrid fiber-reinforced concrete: performance, crack resistance mechanism, and future innovations. Buildings **15**(8), 1247 (2025)
8. Al-Thawadi, S.: Ureolytic bacteria and calcium carbonate formation as a mech- anism of strength enhancement of sand. J. Adv. Sci. Eng. Res. **1**, 98–114 (2011)

9. Anbu, P., Kang, C.-H., Shin, Y.-J., So, J.-S. (2016). Formations of calcium carbonate minerals by bacteria and its multiple applications. Springerplus 5(1), 1–26. doi:https://doi.org/10.1186/s40064-016-1869-2
10. Araújo, M., Van Vlierberghe, S., Feiteira, J., Graulus, G.-J., Van Tittelboom, K., Martins, J.C., Dubruel, N.P., De Belie: Cross-linkable polyethers as healing/sealing agents for self-healing of cementitious materials. Mater. Des. **98**, 215–222 (2016)
11. Bang, S.S., Galinat, J.K., Ramakrishnan, V.: Calcite precipitation induced by polyurethane-immobilized *bacillus pasteurii*. Enzym. Microb. Technol. **28**(4–5), 404–409 (2001). https://doi.org/10.1016/s0141-0229(00)00348-3
12. Banks, E.D., Taylor, N.M., Gulley, J., Lubbers, B.R., Giarizzo, J.G., Bullen, H.A., Hoehler, T.M., Barton, H.A.: Bacterial calcium carbonate precipitation in cave environments: A function of calcium homeostasis. Geomicrobiol J. **27**(5), 444–454 (2010)
13. Cardoso, R., Pedreira, R., Duarte, S., Monteiro, G., Borges, H., Flores-Colen, I.: Biocementation as rehabilitation technique of porous materials. In: New Approaches to Building Pathology and Durability, pp. 99–120. Springer (2016)
14. Chang, J., Yang, D., Lu, C., Shu, Z., Deng, S., Tan, L., Wen, S., Huang, K., Duan, P.: Application of microbially induced calcium carbonate precipitation (MICP) process in concrete self-healing and environmental restoration to facilitate carbon neutrality: a critical review. Environ. Sci. Pollut. Res. **31**(26), 38083–38098 (2024)
15. Cheng, L., Cord-Ruwisch, R.: In situ soil cementation with ureolytic bacteria by surface percolation. Ecol. Eng. **42**(0), 64–72 (2012). https://doi.org/10.1016/j.ecoleng.2012.01.013
16. Cheng, L., Cord-Ruwisch, R., Shahin, M.A.: Cementation of sand soil by microbially induced calcite precipitation at various degrees of saturation. Can. Geotech. J. **50**(1), 81–90 (2013)
17. De Belie, N., De Muynck, W.: Crack repair in concrete using biodeposition. In: Proceedings of the International Conference on Concrete Repair, Rehabilitation and Retrofitting (ICCRRR), Cape Town, South Africa, pp. 291–292. CRC Press (2008)
18. De Muynck, W., De Belie, N., Verstraete, W.: Microbial carbonate precipitation in construction materials: A review. Ecol. Eng. **36**(2), 118–136 (2010). https://doi.org/10.1016/j.ecoleng.2009.02.006
19. Dong, J., Liu, Y., Yuan, S., Li, K., Zhang, F., Guan, Z., Chai, H.K., Wang, Q.: Mechanical behavior and impact resistance of rubberized concrete enhanced by basalt fiber-epoxy resin composite. Constr. Build. Mater. **435**, 136836 (2024)
20. Georget, E., Kapoor, S., Winter, R., Reineke, K., Song, Y., Callanan, M., Ananta, E., Heinz, V., Mathys, A.: In situ investigation of *Geobacillus stearothermophilus* spore germination and inactivation mechanisms under moderate high pressure. Food Microbiol. **41**, 8–18 (2014)
21. Helal, Z., Salim, H., Ahmad, S.S., Elemam, H., Mohamed, A.I., Elmahdy, M.A.: Sustainable bacteria-based self-healing steel fiber reinforced concrete. Case Stud. Constr. Mater. **20**, e03389 (2024)
22. Hussain, A., Ali, D., Koner, S., Hseu, Z.-Y., Hsu, B.-M.: Microbial induce carbonate precipitation derive bio-concrete formation: A sustainable solution for carbon sequestration and eco-friendly construction. Environ. Res. **270**, 121006 (2025)
23. Jang, D., Park, J., Choi, S., Bang, J., Choi, J., Kim, J., Yang, B., Jeon, H.: Novel approach for crack detections and rapid repairment methods in cement-based self-heating composites for smart infrastructures. Compos. Part B **293**, 112126 (2025)
24. Jena, S., Basa, B., Panda, K.C.: Optimizing concrete characteristics with *Bacillus cohnii*: an RSM-based study. J. Build. Pathol. Rehabil. **10**(2), 1–24 (2025)
25. Jonkers, H., Schlangen, E.: Self-healing of cracked concrete: a bacterial approach. In: Proceedings of FRACOS6: Fracture Mechanics of Concrete and Concrete Structures, Catania, Italy, pp. 1821–1826 (2007a)
26. Jonkers, H., Schlangen, E.: Crack repair by concrete-immobilized bacteria. In: Proceedings of the First International Conference on Self Healing Materials, pp. 1–7 (2007b)
27. Jonkers, H., Schlangen, E.: Development of a bacteria-based self healing concrete. In: Proceedings International FIB Symposium, vol. 1, pp. 425–430. CRC Press (2008). https://doi.org/10.1201/9781439828410.ch72

28. Jonkers, H., Thijssen, A., Muyzer, G., Copuroglu, O., Schlangen, E.: Application of bacteria as self-healing agent for the development of sustainable concrete. Ecol. Eng. **36**(2), 230–235 (2010). https://doi.org/10.1016/j.ecoleng.2008.12.036
29. Jonkers, H. (2011). Bacteria-based self-healing concrete. Heron 56 (1/2)
30. Khaliq, W., Ehsan, M.B. (2016). Crack healing in concrete using various bio influenced self-healing techniques. Constr. Build. Mater. **102**, Part 1, 349–357. doi:https://doi.org/10.1016/j.conbuildmat.2015.11.006
31. Kouser, T., Nur, F., Aliyu, A., Altaf, F., Zulfiqar, H., Alhems, L.M.: Applications of epoxy resins as a coating technology in fluid systems. ChemBioEng Rev. **12**(2), e70000 (2025)
32. Krishnapriya, S., Venkatesh Babu, D.L., Prince Arulraj, G.: Isolation and identification of bacteria to improve the strength of the strength of concrete. Microbiol. Res. **174**(0), 48–55 (2015). https://doi.org/10.1016/j.micres.2015.03.009
33. Kumar, M. (2021). A Study of the Abiotic Factors Influencing the Biofilm and Spore Formation of Dairy Isolates of Geobacillus Stearothermophilus and Characterisation of Spores Based on their Heat Resistance: A Thesis Presented in Partial Fulfilment of the Requirements for the Degree of Doctor of Philosophy in Food Safety & Microbiology at Massey University, Palmerston North, New Zealand Massey University
34. Leng, G.-Y., Yan, W., Ye, H.-M., Yao, E.-D., Duan, J.-B., Xu, Z.-X., Li, K.-P., Zhang, J.-R., Li, Z.: Evaluation of the injection and plugging ability of a novel epoxy resin in cement cracks. Pet. Sci. **21**(2), 1211–1220 (2024)
35. López, J.A., Naranjo, J.M., Higuita, J.C., Cubitto, M.A., Cardona, C.A., Vil- lar, M.A.: Biosynthesis of phb from a new isolated bacillus mega- terium strain: outlook on future developments with endospore forming bacteria. Biotechnol. Bioprocess Eng. **17**(2), 250–258 (2012)
36. McMullan, G., Christie, J., Rahman, T., Banat, I., Ternan, N., Marchant, R.: Habitat, applications and genomics of the aerobic, thermophilic genus *Geobacillus*. Biochem. Soc. Trans. **32**(2), 214–217 (2004)
37. Mu, B., Gui, Z., Lu, F., Petropoulos, E., Yu, Y.: Microbial-induced carbonate precipitation improves physical and structural properties of Nanjing Ancient City walls. Materials (Basel). **14**(19), 5665 (2021). https://doi.org/10.3390/ma14195665
38. Onyelowe, K.C., Adam, A.F., Ulloa, N., Garcia, C., Andrade Valle, A.I., Zúñiga Rodríguez, M.G., Zarate Villacres, A.N., Shakeri, J., Anyaogu, L., Alimoradijazi, M.: Modeling the influence of bacteria concentration on the mechanical properties of self-healing concrete (SHC) for sustainable bio-concrete structures. Sci. Rep. **14**(1), 8414 (2024)
39. Otlewska, A., Gutarowska, B.: Environmental parameters conditioning microbially induced mineralisation under the experimental model conditions. Acta Biochim. Pol. **63**, 343–351 (2016)
40. Patel, P. R. And Patel, S. K. (2011). Microbial concrete: the pioneering work on repairing concrete. Int. J. Pharm. Sci. Res. **2**(4), 825–828. https://doi.org/10.13040/IJPSR.0975-8232.2(4).825-28
41. Rahoui, H., Maruyama, I., Vandamme, M., Pereira, J.-M., Mosquet, M.: Impact of alkalis and shrinkage-reducing admixtures on hydration and pore structure of hardened cement pastes. Cem. Concr. Res. **184**, 107620 (2024)
42. Reddy, S.S.P., Rao, M.S., Aparna, P., Sasikala, C.: Performance of ordinary grade bacterial (*Bacillus subtilis*) concrete. Int. J. Earth Sci. Eng. **3**(1), 116–124 (2010)
43. Sharma, N., Ahlawat, Y.K., Mehmood, S., Sharma, A.J., Thakur, P., Malik, A., Sindhu, M., Samani, M., Ahlawat, U.: Industrially significant extremophiles and their unexplored habitats in hot Water Springs. Geomicrobiol J. **42**(1), 40–52 (2025)
44. Van Breugel, K.: Self-healing material concepts as solution for aging infrastructure. In: 37th Conference on our World in Concrete & Structures, Singapore (2012)
45. Van Tittelboom, K., De Belie, N., De Muynck, W., Verstraete, W.: Use of bacteria to repair cracks in concrete. Cem. Concr. Res. **40**(1), 157–166 (2010). https://doi.org/10.1016/j.cemconres.2009.08.025
46. Vishal, A., Chepuri, A., Chandana, N.: Assessment of bacteria-based self-healing concrete through experimental investigations—a sustainable approach. Journal of Materials Science: Materials in Engineering. **20**(1), 15 (2025)

47. Wang, J., Yao, H., Hu, J., Ma, Y., Wang, J.: Dual-encoder network for pavement concrete crack segmentation with multi-stage supervision. Autom. Constr. **169**, 105884 (2025)
48. Wang, J., Snoeck, D., Van Vlierberghe, S., Verstraete, W., De Belie, N.: Application of hydrogel encapsulated carbonate precipitating bacteria for approaching a realistic self-healing in concrete. Constr. Build. Mater. **68**, 110–119 (2014)
49. Wang, J., Soens, H., Verstraete, W., De Belie, N.: Self-healing concrete by use of microencapsulated bacterial spores. Cem. Concr. Res. **56**, 139–152 (2014). https://doi.org/10.1016/j.cemconres.2013.11.009
50. Wang, J., De Belie, N., Verstraete, W.: Diatomaceous earth as a protective vehicle for bacteria applied for self-healing concrete. J. Ind. Microbiol. Biotechnol. **39**(4), 567–577 (2012). https://doi.org/10.1007/s10295-011-1037-1
51. Watanabe, T., Furukawa, S., Hirata, J., Koyama, T., Ogihara, H., Yamasaki, M.: Inactivation of *Geobacillus stearothermophilus* spores by high-pressure carbon dioxide treatment. Appl. Environ. Microbiol. **69**(12), 7124–7129 (2003)
52. Zeigler, D.R.: The *Geobacillus* paradox: why is a thermophilic bacterial genus so prevalent on a mesophilic planet? Microbiology. **160**(Part 1), 1–11 (2014). https://doi.org/10.1099/mic.0.071696-0

# Chapter 2
# Mechanisms of Concrete Self-Healing Derived from Microbiologically Induced Calcite Precipitation (MICP)

## 2.1 Introduction

This chapter explores the mechanisms of self-healing in concrete through Microbiologically Induced Calcite Precipitation (MICP), detailing the chemical reactions involved in the biomineralisation process. It examines the sequence of microbial reactions that lead to the formation of mineral deposits, ultimately enhancing the durability and longevity of concrete. The chapter begins by outlining the fundamental principles of biomineralisation that drive calcite precipitation and contribute to the healing of cracks. It also explains the biological mechanisms underlying the formation and development of mineralised structures. A wide range of bacterial species previously investigated for this purpose is discussed. Finally, the chapter analyses the various factors influencing the biomineralisation process, with a particular focus on the mechanisms involved in crack filling.

## 2.2 Biomineralisation

Biomineralisation is described as the precipitation process of mineral materials influenced by living forms [93]. Biomineralisation refers to the process of mineral formation by living organism which is a widespread phenomenon in nature. Biomineralisation can be accomplished through biologically induced mineralisation process. Biomineralisation can be accomplished through biologically induced mineralisation process. Biologically induces mineralisation usually occurs in an open environment as an uncontrolled consequence of microbial metabolic activity [110]. In this process, biominerals are formed through reaction of metabolic products generated microorganism. Sarayu et al. [94] described biomineralisation as synthesis process under biological mode, and the phase properties like shape, size, crystalline nature and elemental composition are distinguishable due to controlled conditions for the

 
R. M. A. M. Azam and D. H. M. Saman, *Self-Healing Concrete*, Engineering Materials,
https://doi.org/10.1007/978-981-95-7553-4_2

minerals. Biomineralisation creates a local micro-environment that allows chemical precipitation such as calcium carbonate ($CaCO_3$) or also known as calcite.

Biomineralisation process as defined by Anbu et al. [13] and Ojha et al. [73] can occur by three (3) different conditions; biologically controlled mineralisation (BCM), biologically influenced mineralisation (BIfM) and biologically induced mineralisation (BIcM). According to Lowenstam [63], BCM defined as mediated mineralisation of organic matrix performed by many animals, in which the organism control the whole process. For example, the organisms nucleation, growth, morphology, and final location of the deposited mineral depend on its cellular activities. The body synthesises minerals in a form that is unique to that species, independently by environmental conditions [31]. The organisms have a significant degree of control with an isolated occurrence of mineralisation process resulting sophisticated by-products with specific biological functions [105]. However, different mechanisms occurred in BI*f*M in which the presence of cell surfaces of organic matter influenced the process of mineral precipitation [13, 17]. On the other side, the precipitation in BIcM occurs at the cell surface of bacterial species by extracellularly enacted of the cell. In this process, the type of mineral produced is mainly dependent on the environmental conditions [95, 110]. These means the mineral produced by organism would differ depending on the nutrient availability in its surrounding.

Among widespread production of minerals through biomineralisation, precipitation of calcium carbonate has drawn interest due to the efficient bonding capacity and compatibility with concrete application. Precipitation of calcium carbonate in natural environment is ideally influenced by concentration of calcium ions, pH of the solution, concentration of dissolved inorganic carbon and availability of nucleation sites. Bacterial precipitation may be achieved through different pathways like conversion of calcium compound such as Ca-lactate or hydrolysis of urea by ureolytic bacterial metabolism.

Among these three (3) conditions of biomineralisation, both BCM and BI*f*M.

activities controlled by organism primarily. While BI*c*M on another path; the organism has very little control of the process [23] which the mineral produced rigorously manipulated depending on the feeding solution given. Even those three (3) conditions were perceived, the similarities in these three (3) ways of biomineralisation condition are the formation of nucleation sites for crystal growth from organic components which binding a few ions [46, 62, 111].

The most common mineral precipitated in biomineralisation process is the crystal of calcium carbonate ($CaCO_3$) or calcite. However, the bacteria would produce the desired mineral accordingly by feeding the bacteria with a particular feeding solution or depending on availability of the nutrient in its environment. For instance, the sample collected from metal contaminated lake sediment found with the accumulations of complex silicate on the surfaces of bacterial cells [38, 102]. The silicate precipitation was due to the ability of the bacterial species to dissolve silicic acid from metal complexes by the cells and directly precipitate silicate from it [64, 76]. According to Sriaporn et al. [92], the bacterial cells would provide diagenetic silicification sites on relatively high concentrations by the accumulation of organic matter. It was proven in a research study conducted by Inagaki et al. [51] and Afifudin et al.

[8], in which the bacteria fed with silica solution has successfully produced silicate instead of calcite.

In concrete work related to bacteria, the mechanism of mineralisation was based on the understanding of the resulted metabolic by-product mainly due to the nutrient supply under a favourable condition so that the organisms able to create a local microenvironment that allows precipitation of mineral phases such as $CaCO_3$ which defined by BI*c*M. Many kinds of research conducted in self-healing concrete related to bacteria emphasise BI*c*M as a microbial activity that enables organisms to precipitate calcium carbonate/calcite with different names to define the process. For instance, microbially induced calcium carbonate precipitation (MICCP) [80, 87], microbially induced calcite precipitation (MICP) [13, 18, 75], microbially induced precipitation (MIP) [95], microbiological calcium carbonate precipitation (MCP) [31, 34, 79, 106] and calcium precipitation activity (CPA) [9, 112]. MICP will be used to define the biomineralisation process and will be further elaborated in the following sub-section.

Biomineralisation process occurs due to microbial mineral precipitation seeking researchers interest in finding a sustainable yet natural method to improve the behaviour of concrete [7–9, 16, 44, 52, 80, 87, 103]. The biomineralisation process produces calcium carbonate by metabolic activities such as photosynthesis, sulphate reduction and urea hydrolysis [88]. The calcite ($CaCO_3$) forms oolitic and considerable limestone in sediments [10], making it suitable to be explored as an alternative in crack remediation. This method transcends other repair techniques as it is bio-based, profitable, durable and eco-friendly [58]. So far, there are three (3) standard methods of applying MICP in concrete crack remediation namely direct incorporation, surface treatment and also encapsulation. These methods will be elaborated in Chap. 3.

Urease positive bacteria have been found to influence the precipitation of calcium carbonate (calcite) by the production of urease enzyme. This enzyme catalyses the hydrolysis of urea to $CO_2$ and ammonia, resulting in an increase of the pH and calcite precipitation in the bacterial environment [37, 91]. Bacteria produce urease which catalyses the hydrolysis of urea ($CO(NH_2)_2$ into ammonium ($NH^{4+}$) and carbonate ($CO_3{}^{2-}$). First, 1 mol of urea is hydrolysed intracellular to 1 mol of carbonate and 1 mol of ammonia (Eq. 2.1). Carbamate spontaneously hydrolyses to form additionally 1 mol of ammonia and carbonic acid (Eq. 2.2). These products subsequently from 1 mol of bicarbonate and 2 mol of ammonium and hydroxide ions (Eqs. 2.3 and 2.4). The last 2 reactions give rise to a pH increase, which in turn shifts the bicarbonate equilibrium, resulting in the formation of carbonate ions (Eq. 2.5) [33]. The process of making urease for the hydrolysis of urea ($CO(NH_2)_2$ into carbonate $CO_3{}^{2-}$ and ammonium ($NH^{4+}$) is also described by [99]):

$$CO(NH_2)_2 + H_2O \rightarrow NH_2COOH + NH_3 \tag{2.1}$$

$$NH_2COOH + H_2O \rightarrow NH_3 + H_2CO_3 \tag{2.2}$$

$$H_2CO_3 \rightarrow HCO_3^- + H^+ \tag{2.3}$$

$$2NH_3 + 2H_2O \leftrightarrow 2NH^{4+} + 2OH^- \tag{2.4}$$

$$HCO_3^- + H^+ + 2NH^{4+} + 2OH^- \leftrightarrow CO_3^{2-} + 2NH_4^+ + 2H_2O \tag{2.5}$$

Since the cell wall of the bacteria is negatively charged, the bacteria draw cations the environment, including $Ca^{2+,}$ to deposit on their cell surface. The $Ca^{2+}$ ions subsequently react with the $CO_3{}^{2-}$ ions, leading to the precipitation of $CaCO_3$ at the cell surface that serves as a nucleation site (Eqs. 2.6 and 2.7):

$$Ca^{2+} + Cell \leftrightarrow Cell - Ca^{2+} \tag{2.6}$$

$$Cell - Ca^{2+} + CO_3^{2-} \leftrightarrow Cell - CaCO_3 \tag{2.7}$$

Seifan et al. [68]. "Bioconcrete: next generation of self-healing concrete", Applied Microbial Biotechnol, 100:2591–2602.

## 2.3 Microbially Induced Calcite Precipitation (MICP)

Microbially induced calcite precipitation (MICP) refers to the bacterial process that produces calcium carbonate ($CaCO_3$) or calcite as by-product [23]. The evidence of microbial involvement in precipitating $CaCO_3$ from urease activity and its potential has led to the exploration of this method in the multi-field application. For instance, MICP has been investigated for treatment of radionuclide and remediation of heavy metals [28, 41, 48, 101], sand consolidation [1], sand cementation [26, 27], soil improvement [70] and ornamental and monumental stone conversation [55, 84, 97]. The microbial precipitation of $CaCO_3$ is determined by several factors including: the concentration of dissolved inorganic carbon, the pH, the concentration of calcium ions and the presence of nucleation sites. The first three factors are provided by the metabolism of the bacteria while the cell wall of the bacteria will act as a nucleation sites [49].

Wide range of microorganisms has the ability to perform urease activity with some strains producing high levels of urease. High precipitation amount of $CaCO_3$ by using microorganisms with high urease activity [91] relatively influenced the selection of a specific strain to be used to serve the purpose. The common type of bacteria used for the production of urease and $CaCO_3$ are from genus *Bacillus* [30, 52, 58, 78, 83, 103, 104]. *Sporosarcina pasteurii* (formerly known as *Bacillus pasteurii*) is one of the bacterial strain that has high urease activity [13], 30, 33]. Due to this fact, many research conducted using *Sporosarcina pasteurii* for MICP have been reported [2, 4, 15, 19, 23, 29, 65, 78, 82, 91, 92]. Other bacteria that have the ability to produce urease activity in genus *Bacillus* also have been studied such as *Bacillus aerius* [89], *Bacillus subtilis* [57, 83]. *Bacillus pseudofirmus* [52, 112], *Bacillus Megaterium* [14,

58, 61], *Bacillus licheniformis* and *Bacillus flexus* [58], *Bacillus sphaericus* [30, 85, 86, 99, 103, 104] and *Bacillus cohnii* [53] has been reported being used in MICP.

The most common and well-known pathways to precipitate $CaCO_3$ is by enzymatic hydrolysis of urea ($CO(NH_2)^2$) from bacteria through biomineralisation process [13, 31, 75]. Urea hydrolysis can produce high concentrations of carbonate within a short period and is the most easily controlled reaction to generate carbonate [50, 106]. The equation for this chemical reaction shown in Eq. 2.8.

$$CO(NH_2)_2 + 2H_2O \rightarrow 2NH^+ + CO^{2-} \tag{2.8}$$

Urea hydrolysis process from urease activity resulting pH rises by the released of ammonium ($NH^{4+}$) thus promotes the precipitation of calcium carbonate ($CaCO_3$). The combination of calcium ($Ca^{2+}$) from the feeding solution (nutrient) or environment with carbonate ions ($CO^{2-}$) resulting from the formation of calcium carbonate ($CaCO_3$) or calcite at the cell surface that serves as nucleation sites for bacteria [23] as shown in Eq. 2.9.

$$Ca^{2+} + CO^{2-} \rightarrow CaCO_3 \tag{2.9}$$

Many types of feeding solution (nutrient) reported in the literature. The overview of the different nutrients to promote the formation of calcium carbonate resulting from MICP process shown as in Table 2.1.

## 2.4 Factors Influencing Microbially Induced Calcite Precipitation (MICP)

The basis of bacterial to perform urea hydrolysis and calcite precipitation depends on the biomineralisation capability driven by the bacteria. Tobler et al. [98] derived the kinetics of bacterial urea hydrolysis and calcite precipitation by relating the rates of bacterial urea hydrolysis ($k_{urea}$), calcite precipitation ($k_p$) and calcite saturation ($k_s$). The rate of urea hydrolysis and calcite saturation will influence the rate of calcite precipitation of ureolytic bacteria. As proposed by Tobler et al. [98], to promote ureolysis-driven calcite precipitation, sufficient organisms with ureolytic activity should be present in the microbial community.

The biomineralisation grasped on the ability of ureolytic bacteria for urea hydrolysis and calcite precipitation development on its basis of kinetics. A limited study performed on ureolysis rate investigation as in-situ. It is because the microorganism size and in-situ environment are difficult to be assessed and accessed. Few studies determined ureolysis rates by ureolytic bacteria as in-situ natural groundwater and natural soil environment [42]. Therefore, the mimicked environment or laboratory simulation created as such to investigate and predict the ureolytic rate by bacteria to represent and model the real condition of bacteria application.

**Table 2.1** The overview of the different feeding solutions (nutrients) used for microbially induced calcite precipitation (MICP)

| Name | Composition | | Cell concentrations | pH | Temperature | References |
|---|---|---|---|---|---|---|
| | Nutrients | Concentration | | | | |
| Media composition | Yeast extract | 5 g/l | $10^5$ cfu/ml | – | 37 °C | Krishnapriya et al. [58] |
| | Beef extract | 5 g/l | | | | |
| | Wheat bran | 20 g/l | | | | |
| Bio-reagants | Yeast extract | 40 g/l | $10^9$ cfu/ml | 9.0 | – | Wang et al. [104] |
| | Urea | 90 g/l | | | | |
| | Calcium nitrate | 120 g/l | | | | |
| Precipitation agar | Urea | 20 g/l | | – | 28 °C | Chahal et al. [19] |
| | $NH_4Cl$ | 2.12 g/l | $10^3$ cfu/ml | | | |
| | $NaHCO_3$ | 10 g/l | $10^5$ cfu/ml | | | |
| | $CaCl_2.2H_2O$ | 25 g/l | $10^7$ cfu/ml | | | |
| | Nutrient broth | 3 g/l | | | | |
| Liquid medium | Bacto | 3 g/l | – | 6.0–9.0 | 30 °C | Park and Lee [77] |
| | Urea | 20 g/l | | | | |
| | NH4Cl | 10 g/l | | | | |
| | $NaHCO_3$ | 2.12 g/l | | | | |
| | $CaCl_2.2H_2O$ | 3.7 g/l | | | | |
| | Agar powder | 15 g/l | | | | |
| Medium composition | Peptone | 5 g/lt | $10^5$ cfu/ml | - | 37 °C | Reddy et al. [83] |
| | NaCl | 5 g/lt | | | | |
| | Yeast extract | 3 g/lt | | | | |

The laboratory simulation by artificial groundwater has been performed using a urease positive strain of *Sporosarcina pasteurii* [35, 39, 41, 65]. The objective of the simulation focused on temperature variability, the size of inoculum or different $Ca^{2+}$ and urea concentration effect on the mechanisms and kinetics of ureolysis rates and calcite precipitation. Dupraz et al. [35] summarised the kinetic parameters of urea hydrolysis ($k_{urea}$), mostly affected by temperature. Reported $k_{urea}$ values range between 0.05 $day^{-1}$ to 0.9 $day^{-1}$ occurred due to the applied $Ca^{2+}$ concentrations 1.75 mM to 20 mM, the quantities of calcite that produced in these experiments were relatively small, 0.2 to 2 g $L^{-1}$ $CaCO_3$.

The effects of bacterial cell concentration from *Sporocarsina pasteurii*, urea concentration and $Ca^{2+}$ concentration on the values of mean $k_{urea}$ and mean mass $CaCO_3$ precipitation was inferred by Okwadha and Li [74] as the increased value of $k_{urea}$ has a proportional effect on the increasing number of bacteria cell concentration [54]. By increasing the cell concentration, the nucleation sites' number would lead to increase that corresponds to the higher production of calcite [32, 40]. With such results, the optimum condition by higher exposure $Ca^{2+}$ is also one of the contributors. The optimal conditions for the MICP to occur were identified, at a bacterial cell concentration of $2.3 \times 10^8$ cells $mL^{-1}$ which exposed to 666 mM urea concentration and 250 mM calcium concentration. In this finding, the mean mass $CaCO_3$ precipitation under such conditions is the highest ($13.0 \pm 1.30$ mg $d^{-1}$) in comparison to the values under other conditions evaluated in the experimental work [107]. Wong [107] asserted a similar evolutionary pattern from the research work of Tobler et al. [98] in the process of MICP that correlate $k_{urea}$ value with bacterial cell concentration. Based on study conducted by Tobler et al. [98], using the same urea concentration at 250 mM, with 200 mM of $Ca^{2+}$ concentration and temperature of 20 °C, with the increasing number of bacterial cell concentration from $7.2 \times 10^5$ cells $mL^{-1}$ to $2.3 \times 10^6$ cells $mL^{-1}$, the respective $k_{urea}$, $k_s$ and $k_p$ values of the bacteria drastically increased by 94.3, 78.6 and 82.9%.

Also, it is noticeable that by increasing the $Ca^{2+}$ concentration from 50 to 200 mM at the same bacterial cell concentration of $7.2 \times 10^5$ cells $mL^{-1}$, urea concentration of 250 mM and temperature of 20 °C, the changes of $k_{urea}$, $k_s$ and $k_p$ values of the bacteria are not obvious although the $k_p$ value demonstrates an increasing trend with $Ca^{2+}$ concentration. Higher $k_{urea}$ value is expounded by higher bacterial concentration. It has a direct and positive influence on MICP. Urease catalyses the hydrolysis of urea into ammonium ($NH^+$) and carbonate ($CO_2^-$) [100]. The higher amount of urease enzyme produced by higher bacterial concentration would escalate chemical conversion of urea into $NH^{4+}$ and $CO_3^{2-}$. The subsequent increase in pH due to the presence of ammonia ions and the additional release of $CO_2$ from the enzymatic urea hydrolysis accelerate the rate of the urease-induced calcite precipitation [60]. Therefore, higher bacterial concentration plays an important role as the active participation of urease in enhancing the rates of biochemical calcite precipitation.

Based on studies conducted by Ferris et al. [39], and Mitchell and Ferris [65], the temperature has no definitive trend in the rate of calcite precipitation as compared to the rate of urea hydrolysis [107]. Temperature increase does not influence the calcite saturation rate ($k_s$) profoundly, and the calcite precipitation, $k_p$ value unlike the ureolysis rate, $k_{urea}$. Ferris et al. [39] and Mitchell & Ferris [65] setting a condition in which the respective urea and $Ca^{2+}$ concentrations were fixed to 1.75 mm and 6 mm while Tobler et al. [98] set those conditions at 200 mm and 250 mm respectively. As predicted, the result obtained by Ferris et al. [39] and Mitchell & Ferris [65] for $k_p$ was $0.07 \times 10^{-1}$ mM $d^{-1}$ while Tobler et al. [98] obtained it at 2.43 to $2.50 \times 10^{-1}$ mM $d^{-1}$. These results indicated the limitation of bacteria to metabolise urea due to the lower urea concentration present in the artificial groundwater. Due to the restriction on urea metabolised by bacteria, no pH arises in the system. Thus, the process of calcite precipitation will be retarded.

From studies conducted by Ferris et al., [39]; Mitchell and Ferris [65]; Okhwada and Li [74]; Tobler et al. [98], it was concluded that the incorporation of *Sporocarsina pasteurii* in artificial groundwater has increased the pH of growth media from 6.5 to 9. The slow increase of pH was obtained; resulted from the lower urea concentration that produced less carbonate [98]. Therefore, to achieve optimum calcite precipitation rates in cementitious material, the utilisation of bacteria that able to survive in such alkaline environment would contribute to the high production of carbonate, urea concentration; thus, producing more calcite that beneficial in autonomous healing concrete.

## 2.5 Types of Bacteria Agents Used as Self-Healing Agents in Cementitious Materials

A wide range of bacteria has been reported to produce self-healing concrete [11, 12, 52, 104, 112]. The type of bacteria used also has a direct influence on the effectiveness of the MICP process in concrete. Two (2) major groups of bacteria used in MICP namely mesophile and thermophile [96]. Mesophile is the bacteria that grow best at temperature 20 °C to 45 °C [71]. Thermophilic bacteria are defined as bacteria that thrive in temperature of 45 °C to 70 °C [59].

Harsh environment of concrete due to the pH of the fresh concrete is between 10 and 13, is not suitable for the survival of the majority of bacteria. Also, the temperature of fresh concrete can reach up to 70 °C especially in the core structure is also too high for the cell growth [96]. Therefore, thermophilic bacteria is more suitable to be used as a healing agent in concrete. Meanwhile, further hydration causes limited space for the cell to grow; thus, it diminishes the cell viability. Therefore, the bacteria that suitable to be used in self-healing concrete should exhibit high resistance against high pH, temperature and limitation of water supply. Unlike the vegetative cell, bacterial spore is very resistant against the harsh conditions as mentioned and some bacterial spores can live for more than 60 years [53]. Therefore, the deliberation on the application of bacteria types revolved on bacteria that can produce endospore as a survival strategy to be applied in self-healing concrete.

Thermophilic bacteria was first incorporated into the cement-based material by Ghosh et al. [43]. Thermophilic bacteria can be found in a natural geothermal region such as hot springs and possess high metabolic rates, physically and chemically stable enzymes. Also, thermophilic bacteria have lower growth, but higher end product as a consequence of growth at high temperature with unique macromolecular properties [108]. Ghosh et al. [43] used *Shewanella* at $10^1$ cfu/ml to $10^7$ cfu/ml cell concentrations. Ghosh and his team [43] investigated the effect of including bacteria to the performance of the mortar. The results revealed that the maximum strength obtained by the bacteria inclusion was at $10^5$ cfu/ml concentration. At the same bacterial concentration, the greatest reduction in porosity was found as compared with those made other bacterial concentration. In conjunction with this research, another type of thermophilic anaerobic bacteria isolated from hot spring was used in research conducted by Ghosh et al. [44] and Ghosh et al. [45]. Anaerobic bacteria defined as bacteria that are capable of living in the absence of air or free oxygen [47]. The results revealed that the enrichment culture supplied to the thermophilic bacteria enables the bacteria to grow inside the concrete or mortar matrix leading to the formation of crystals that enhance the refinement of pore structure. The purpose of using anaerobic bacteria was because they would grow well in concrete or mortar matrix without a supply of oxygen and food, able to decrease permeability, modify porosity of the matrices and repair more efficiently the cracks of concrete/mortar samples when incorporated as highlighted by Ghosh et al. [44] and Ghosh et al. [45].

A similar pattern regarding mortar properties enhancement using thermophilic bacteria also discovered by Biswas et al. [16]. Biswas et al. [16] using bacteria isolated from a hot spring, namely *Thermoanaerobactor thermohydrosulfuricus*. The bacterial strain increases the compressive and tensile strengths of both cement paste and mortar through one of its unique secretory proteins (homologous to carbonic anhydrase-II). This technique also has been patented as described in Chattopadhyay and Mandal [24, 25].

The authors used *Geobacillus stearothermophilus* bacteria type that has not yet been explored in the cementitious environment. As *Geobacillus* came from the genus *Bacillus*, it has all the same morphology as its predecessor. *Geobacillus* also defined as earth *Bacillus* because it is easily found in thermal vents and grouped under thermophilic due to its character as 'heat-loving' bacteria. While the species name; stearos (fat) + thermo (heat) + philus (loving) means fat and heat-loving bacterium [109]. *Geobacillus stearothermophilus* is a spore-forming bacteria with growth temperature range of 35 °C to 70 °C. These bacteria also able to lie dormant for extended periods. Spore formation was triggered by nutrient deprived, and unfavourable condition for bacteria lie quiescent. Spores of *Geobacillus stearothermophilus* are the most heat-resistant among spores of aerobic spore-forming bacteria, and can spoil a variety of types of foodstuffs [20–22].

Previously, *Geobacillus stearothermophilus* has been used as an indicator of the primary thermophilic spore former involved in flat sour spoilage of canned foods [36, 69]. To evaluate the effectiveness of sterilisation processes, *Geobacillus stearothermophilus* act as a biological indicator [113]. The potential of thermolysin derived from *Geobacillus stearothermophilus* as a zinc metalloprotease that has potential use in the food industry [56]. Besides, *Geobacillus stearothermophilus* also has been utilised in the biogenic synthesis process of silver and gold nanoparticles [66].

**Table 2.2** Several types of bacteria species used for calcite precipitation in concrete

| Bacteria | References |
|---|---|
| *AKKR5* | Siddique et al. [89] |
| *Bacillus aerius* | Siddique et al. [89]; Siddique et al. [90] |
| *Bacillus megaterium* | Achal, Pan & Özyurt [3]; Andalib et al. [14]; Krishnapriya et al. [58] |
| *Bacillus* sp. *CT5* | Achal et al. [6] |
| *Bacillus subtilis* | Khaliq and Ehsan [57]; Mondal and Ghosh (Dey) [67]; Nosouhian et al. [72]; Sahoo et al. [85] |
| *Sporosarcina pasteurii* | Chahal et al. [19]; Nosouhian et al. [72]; Ramakrishnan et al. [81] |
| *Shewanella species* | Ghosh et al. [43] |
| *Bacillus sphaericus* | [5]; De Myunck et al. [30]; [72]; Van Tittleboom et al. [99] |
| *Bacillus pasteruii* | Ramakrishnan et al. [81]; Van Tittleboom et al. [99] |

Several other microorganisms have the ability to precipitate calcium carbonate by means of urealysis. It has been shown that for instance, *Bacillus Pasteruii*, a common soil bacterium, can continuously precipitate a new highly impermeable calcite layer over the surface of an already existing concrete layer. The *Bacillus Subtilis* bacteria can increase the strength of the concrete with lightweight aggregate and graphite nanoplatelets [57]. *Bacillus aerius* bacteria in rice husk ash concrete was investigated and it was observed that the durability of concrete has been increased [89, 90]. *Bacillus megaterium* bacteria were used in concrete and the results showed a 24% increase in compressive strength [14]. The deposition of calcium carbonate in concrete by *Bacillus Sphaericus* improves the durability of concrete [104]. The *Sporoscarcina pasteurii* bacteria used in fly ash concrete has shown improvement in strength and durability of fly ash concrete through self-healing effect [19].

***Sporosarcina pasteurii*** was incorporated into silica fume concrete, resulting in enhanced strength and durability due to its self-healing capabilities [19]. Similarly, ***Bacillus sphaericus*** was applied to concrete surfaces to evaluate its effectiveness as a surface treatment.

The findings indicated that bacterial-induced carbonate precipitation offers a promising alternative for concrete surface treatment [30]. Table 2.2 presents various types of bacteria that have been employed as self-healing agents or in related applications within concrete and mortar.

## References


1. Abo-El-Enein, S.A., Ali, A.H., Talkhan, F.N., Abdel-Gawwad, H.A.: Utilization of microbial induced calcite precipitation for sand consolidation and mortar crack remediation. HBRC J. **8**(3), 185–192 (2012). https://doi.org/10.1016/j.hbrcj.2013.02.001
2. Achal, V., Pan, X.: Characterization of urease and carbonic anhydrase producing bacteria and their role in calcite precipitation. Curr. Microbiol. **62**(3), 894–902 (2011). https://doi.org/10.1007/s00284-010-9801-4

3. Achal, V., Pan, X., Özyurt, N.: Improved strength and durability of fly ash-amended concrete by microbial calcite precipitation. Ecol. Eng. **37**(4), 554–559 (2011b)
4. Achal, V., Siddique, R., Reddy, M. and Mukherjee, A. (2009) Improvement in the compressive strength of cement mortar by the use of a microorganism—*bacillus megaterium*. In: Limbachiya, Kew Excellence in Concrete Construction through Innovation, pp. 27–29
5. Achal, V., Mukherjee, A., Reddy, M.: Microbial concrete: way to enhance the durability of building structures. J. Mater. Civ. Eng. **23**(6), 730–734 (2011a). https://doi.org/10.1061/(ASCE)MT.1943-5533.0000159
6. Achal, V., Mukerjee, A., Sudhakara Reddy, M.: Biogenic treatment improves the durability and remediates the cracks of concrete structures. Construction Building Materials. **48**, 1–5 (2013). https://doi.org/10.1016/j.conbuildmat.2013.06.061
7. Achal, V., Mukherjee, A., Kumari, D., Zhang, Q.: Biomineralization for sustainable construction—a review of processes and applications. Earth Sci. Rev. **148**, 1–17 (2015). https://doi.org/10.1016/j.earscirev.2015.05.008
8. Afifudin, H., Nadzarah, W., Hamidah, M.S., Noor Hana, H.: Microbial participation in the formation of calcium silicate hydrated (CSH) from *Bacillus subtilis*. Procedia Eng. **20**(0), 159–165 (2011). https://doi.org/10.1016/j.proeng.2011.11.151
9. Ahmad, M.A., Zhang, J., Liu, B., Guohao, X., Xiaoyi, T., Haoying, G., Changjie, S., Runhao, L., Xiaona, X., Weilin, L., Huang, R., Peiwen, T., Deng, X.: Synergistic effect of composite bacteria on self-healing process of concrete crack. Case Stud. Constr. Mater. **20**, e03028 (2024). https://doi.org/10.1016/j.cscm.2024.e03028
10. Al-Thawadi, S.: Ureolytic bacteria and calcium carbonate formation as a mechanism of strength enhancement of sand. Journal of Advanced Science and Engineering Research. **1**, 98–114 (2011)
11. Amirreza, T., Mohamad Ali, F., Ali, K., Ramin, A., Muhd Zaimi, A., Mohanadoss, P., Bin, R., zin, R.M., Lee, C.T., Shafaghat, A., Warid, M., Ir, H., Mohd Warid, H.: Identification of gaps to conduct a study on biological self-healing concrete. J. Environ. Treat. Tech. **1**(2), 62–68 (2013)
12. Annamalai, S.K., Arunachalam, K.D., Sathyanarayanan, K.S.: Production and characterization of bio caulk by *bacillus pasteurii* and its remediation properties with carbon nano tubes on concrete fractures and fissures. Mater. Res. Bull. **47**(11), 3362–3368 (2012). https://doi.org/10.1016/j.materresbull.2012.07.024
13. Anbu, P., Kang, C.-H., Shin, Y.-J., So, J.-S.: Formations of calcium carbonate minerals by bacteria and its multiple applications. Springer Plus. **5**(1), 1–26 (2016). https://doi.org/10.1186/s40064-016-1869-2
14. Andalib, R., Muhd Zaimi, A.M., Mohd Warid, H., Mohanadoss, P., Ali, K., Jahangir, M., Han-Seung, L.: Optimum concentration of bacillus megaterium for strengthening structural concrete. Constr. Build. Mater. **118**, 180–193 (2016). https://doi.org/10.1016/j.conbuildmat.2016.04.142
15. Bang, S.S., Galinat, J.K., Ramakrishnan, V.: Calcite precipitation induced by polyurethane-immobilized *bacillus pasteurii*. Enzym. Microb. Technol. **28**(4–5), 404–409 (2001). https://doi.org/10.1016/s0141-0229(00)00348-3
16. Biswas, M., Majumdar, S., Chowdhury, T., Chattopadhyay, B., Mandal, S., Halder, U., Yamasaki, S.: Bioremediase a unique protein from a novel bacterium BKH1, ushering a new hope in concrete technology. Enzym. Microb. Technol. **46**(7), 581–587 (2010). https://doi.org/10.1016/j.enzmictec.2010.03.005
17. Benzerara, K., Miot, J., Morin, G., Ona-Nguema, G., Skouri-Panet, F., Ferard, C.: Significance, mechanisms and environmental implications of microbial biomineralization. Compt. Rendus Geosci. **343**(2), 160–167 (2011)
18. Bosak, T.: Calcite precipitation, microbially induced. In: Encyclopedia of Geobiology, pp. 223–227. Springer Netherlands (2011). https://doi.org/10.1007/978-1-4020-9212-1_41
19. Chahal, N., Siddique, R., Rajor, A.: Influence of bacteria on the compressive strength, water absorption and rapid chloride permeability of fly ash concrete. Constr. Build. Mater. **28**(1), 351–356 (2012). https://doi.org/10.1016/j.conbuildmat.2011.07.042

20. Feeherry, F.E., Munsey, D.T., Rowley, D.B.: Thermal inactivation and injury of Bacillus stearothermophilus spores. Appl. Environ. Microbiol. **53**(2), 365–370 (1987)
21. Prevost, S., Andre, S., Remize, F.: PCR detection of thermophilic spore-forming bacteria involved in canned food spoilage. Curr. Microbiol. **61**(6), 525–533 (2010)
22. Burgess, S.A., Lindsay, D., Flint, S.H.: Thermophilic bacilli and their importance in dairy processing. Int. J. Food Microbiol. **144**(2), 215–225 (2010)
23. Cardoso, R., Pedreira, R., Duarte, S., Monteiro, G., Borges, H., Flores-Colen, I.: Biocementation as rehabilitation technique of porous materials. In: New Approaches to Building Pathology and Durability, pp. 99–120. Springer (2016)
24. Chattopadhyay, B., Mandal, S. (2013, August). Materials and methods for the production of green concrete. Google patents (US patent 8,518,177)
25. Chattopadhyay, B., Mandal, S. (2014, September). Hot spring bacterial strain bkh1 and protein isolated therefrom, concrete compositions, and uses thereof. Google patents (US patent 8,835,151)
26. Cheng, L., Cord-Ruwisch, R., Shahin, M.A.: Cementation of sand soil by microbially induced calcite precipitation at various degrees of saturation. Can. Geotech. J. **50**(1), 81–90 (2013)
27. Chou, C.-W., Seagren, E.A., Aydilek, A.H., Lai, M.: Biocalcification of sand through ureolysis. J. Geotech. Geoenviron. **137**(12), 1179–1189 (2011)
28. Colwell, F.S., Smith, R.W., Ferris, F.G., Reysenbach, A.-L., Fujita, Y., Tyler, T.L., Taylor, J.L., Banta, A., Delwiche, M.E., McLing, T.L.: Microbially mediated subsurface calcite precipitation for removal of hazardous divalent cations: microbial activity, molecular biology, and modeling. In. ACS Publications (2005)
29. Cuthbert, M.O., Riley, M.S., Handley-Sidhu, S., Renshaw, J.C., Tobler, D.J., Phoenix, V.R., Mackay, R.: Controls on the rate of ureolysis and the morphology of carbonate precipitated by *s. pasteurii* biofilms and limits due to bacterial encapsulation. Ecol. Eng. **41**, 32–40 (2012). https://doi.org/10.1016/j.ecoleng.2012.01.008
30. De Muynck, W., Cox, K., De Belie, N., Verstraete, W.: Bacterial carbonate precipitation as an alternative surface treatment for concrete. Constr. Build. Mater. **22**(5), 875–885 (2008). https://doi.org/10.1016/j.conbuildmat.2006.12.011
31. De Muynck, W., De Belie, N., Verstraete, W.: Microbial carbonate precipitation in construction materials: A review. Ecol. Eng. **36**(2), 118–136 (2010). https://doi.org/10.1016/j.ecoleng.2009.02.006
32. Dhami, N.K., Mukherjee, A., Reddy, S.M., Seto, J.: Biofilm and Microbial Applications in Biomineralized Concrete. INTECH Open Access Publisher (2012)
33. Dick, J., De Windt, W., De Graef, B., Saveyn, H., Van der Meeren, P., De Belie, N., Verstraete, W.: Bio-deposition of a calcium carbonate layer on degraded limestone by *bacillus* species. Biodegradation. **17**(4), 357–367 (2006). https://doi.org/10.1007/s10532-005-9006-x
34. Dung, N.T., Hoang, T., Yang, E.-H., Chu, J., Unluer, C.: New frontiers in sustainable cements: improving the performance of carbonated reactive MgO concrete via microbial carbonation process. Constr. Build. Mater. **356**, 129243 (2022). https://doi.org/10.1016/j.conbuildmat.2022.129243
35. Dupraz, S., Parmentier, M., Ménez, B., Guyot, F.: Experimental and numerical modeling of bacterially induced pH increase and calcite precipitation in saline aquifers. Chem. Geol. **265**(1), 44–53 (2009)
36. Durand, L., Planchon, S., Guinebretiere, M.-H., Carlin, F., Remize, F.: Genotypic and phenotypic characterization of foodborne *Geobacillus stearothermophilus*. Food Microbiol. 45(0), Part A, 103–110 (2015). https://doi.org/10.1016/j.fm.2014.01.015
37. Feng, J., Dang, W., Gao, Q., Jiang, K., Zhu, S., Ni, J., Zhang, J., Lu, P., Wei, W., Song, H., Liu, X., Wang, Z.: Urea-hydrolyzing bacillus pasteurii microbially induced calcite precipitation for Ca2+ removal and enhanced anaerobic granular sludge activity in high-calcium wastewater. J. Water Process Eng. **68**, 106545 (2024). https://doi.org/10.1016/j.jwpe.2024.106545
38. Ferris, F., Fyfe, W., Beveridge, T.: Bacteria as nucleation sites for authigenic minerals in a metal-contaminated lake sediment. Chem. Geol. **63**(3–4), 225–232 (1987)

39. Ferris, F., Phoenix, V., Fujita, Y., Smith, R.: Kinetics of calcite precipitation induced by ureolytic bacteria at 10 to 20°C in artificial groundwater. Geochim. Cosmochim. Acta. **68**(8), 1701–1710 (2004)
40. Fortin, D., Ferris, F., Beveridge, T.: Surface-mediated mineral development by bacteria. Rev. Mineral. Geochem. **35**(1), 161–180 (1997)
41. Fujita, Y., Redden, G.D., Ingram, J.C., Cortez, M.M., Ferris, F.G., Smith, R.W.: Strontium incorporation into calcite generated by bacterial ureolysis. Geochim. Cosmochim. Acta. **68**(15), 3261–3270 (2004)
42. Fujita, Y., Taylor, J.L., Gresham, T.L., Delwiche, M.E., Colwell, F.S., McLing, T.L., Petzke, L.M., Smith, R.W.: Stimulation of microbial urea hydrolysis in ground- water to enhance calcite precipitation. Environ. Sci. Technol. **42**(8), 3025–3032 (2008)
43. Ghosh, P., Mandal, S., Chattopadhyay, B.D., Pal, S.: Use of microorganism to improve the strength of cement mortar. Cem. Concr. Res. **35**(10), 1980–1983 (2005). https://doi.org/10.1016/j.cemconres.2005.03.005
44. Ghosh, P., Mandal, S., Chattopadhyay, B.D.: Effect of addition of microorganisms on the strength of concrete. Indian Concrete Journal. **80**(4), 45–48 (2006)
45. Ghosh, P., Mandal, S., Pal, S., Bandyopadhyaya, G., Chattopadhyay, B.D.: Development of bioconcrete material using an enrichment culture of novel thermophilic anaerobic bacteria. Indian J. Exp. Biol. **44**(4), 336 (2006c)
46. Gotliv, B.-A., Kessler, N., Sumerel, J.L., Morse, D.E., Tuross, N., Addadi, L., Weiner, S.: Asprich: A novel aspartic acid-rich protein family from the prismatic shell matrix of the bivalve Atrina rigida. Chembiochem. **6**(2), 304–314 (2005)
47. Green, S.: A definition and systems view of anaerobic capacity. Eur. J. Appl. Physiol. Occup. Physiol. **69**(2), 168–173 (1994)
48. Hamdan, N., Kavazanjian Jr, E., Rittmann, B.E.: Sequestration of radionuclides and metal contaminants through microbially-induced carbonate precipitation. Proceedings 14th Pan American Conference Soil Mechanic. Geotechnical Engineering, Toronto, Canada (2011)
49. Hammes, F., Boon, N., de Villiers, J., Verstraete, W., Siciliano, S.D.: Strain-Specific ureolytic microbial calcium carbonate precipitation. Appl. Environ. Microbiol. **69**(8), 4901–4909 (2003). https://doi.org/10.1128/AEM.69.8.4901-4909.2003
50. Huang, H., Ye, G., Qian, C., Schlangen, E.: Self-healing in cementitious materials: materials, methods and service conditions. Mater. Des. **92**, 499–511 (2016)
51. Inagaki, F., Yokoyama, T., Doi, K., Izawa, E., Ogata, S.: Bio-deposition of amorphous silica by an extremely thermophilic bacterium, *Thermus spp.* Biosci. Biotechnol. Biochem. **62**(6), 1271–1272 (1998)
52. Jonkers, H., Schlangen, E. (2007). Self-healing of cracked concrete: a bacterial approach. In: Proceedings of FRACOS6: Fracture Mechanics of Concrete and Concrete Structures, Catania, Italy, pp. 1821–1826
53. Jonkers, H., Schlangen, E. (2008) Development of a bacteria-based self-healing concrete. In: Proceedings International FIB Symposium, vol. 1, 425–430
54. Jonkheijm, P., van der Schoot, P., Schenning, A.P., Meijer, E.: Probing the solvent-assisted nucleation pathway in chemical self-assembly. Science. **313**(5783), 80–83 (2006)
55. Jiménez-López, C., Rodriguez-Navarro, C., Piñar, G., Carrillo-Rosúa, F., Rodriguez- Gallego, M., Gonzalez-Muñoz, M.: Consolidation of degraded ornamental porous limestone stone by calcium carbonate precipitation induced by the microbiota inhabiting the stone. Chemosphere. **68**(10), 1929–1936 (2007)
56. Ke, Q., Chen, A., Minoda, M., Yoshida, H.: Safety evaluation of a thermolysin enzyme produced from *Geobacillus stearothermophilus*. Food Chem. Toxicol. **59**(0), 541–548 (2013). https://doi.org/10.1016/j.fct.2013.06.046
57. Khaliq, W., Ehsan, M.B.: Crack healing in concrete using various bio influenced self-healing techniques. Constr. Build. Mater. **102**, Part 1, 349–357 (2016). https://doi.org/10.1016/j.conbuildmat.2015.11.006
58. Krishnapriya, S., Venkatesh Babu, D.L., Prince Arulraj, G.: Isolation and identification of bacteria to improve the strength of concrete. Microbiol. Res. **174**(0), 48–55 (2015). https://doi.org/10.1016/j.micres.2015.03.009

59. Kristjansson, J.K.: Thermophilic bacteria. In: Kristjansson, J.K. (ed.) Technological Institute of Iceland and University of Iceland, Reykjavik. CRC press LLC, Iceland (1991)
60. Li, H., Song, Y., Li, Q., He, J., Song, Y.: Effective microbial calcite precipitation by a new mutant and precipitating regulation of extracellularurease. Bioresour. Technol. **167**, 269–275 (2014). https://doi.org/10.1016/j.biortech.2014.06.011
61. Lian, B., Hu, Q., Chen, J., Ji, J., Teng, H.H.: Carbonate biomineralization induced by soil *bacterium bacillus megaterium*. Geochim. Cosmochim. Acta. **70**(22), 5522–5535 (2006). https://doi.org/10.1016/j.gca.2006.08.044
62. Lowenstam, H., Weiner, S.: On biomineralization. Oxford University Press, New York (1988)
63. Lowenstam, H.: Minerals formed by organisms. Science. **211**(4487), 1126–1131 (1981)
64. McCutchin, C.A., Edgar, K.J., Chen, C.-L., Dove, P.M.: Silica–Biomacromolecule Interactions: Toward a Mechanistic Understanding of Silicification. Biomacromolecules. **26**(1), 43–84 (2024)
65. Mitchell, A.C., Ferris, F.G.: The coprecipitation of Sr into calcite precipitates induced by bacterial ureolysis in artificial groundwater: temperature and kinetic dependence. Geochim. Cosmochim. Acta. **69**(17), 4199–4210 (2005)
66. Mohammed Fayaz, A., Girilal, M., Rahman, M., Venkatesan, R., Kalaichelvan, P.T.: Biosynthesis of silver and gold nanoparticles using thermophilic bacterium *Geobacillus stearothermophilus*. Process Biochem. **46**(10), 1958–1962 (2011). https://doi.org/10.1016/j.procbio.2011.07.003
67. Mondal, S., Ghosh A (Dey).: Spore-forming *Bacillus subtilis* Vis-à-Vis non-spore-forming *Deinococcus radiodurans*, a novel bacterium for self-healing of concrete structures: a comparative study. Constr. Build. Mater. 266, Part B (2021). https://doi.org/10.1016/j.conbuildmat.2020.121122
68. Seifan, M., Samani, A.K., Berenjian, A.: Bioconcrete: next generation of self-healing concrete. Appl. Microb. Biotechnol. **100**, 2591–2602 (2016)
69. Mtimet, N., Trunet, C., Mathot, A.-G., Venaille, L., Leguérinel, I., Coroller, L., Couvert, O.: Modeling the behavior of Geobacillus stearothermophilus ATCC 12980 throughout its life cycle as vegetative cells or spores using growth boundaries. Food Microbiol. **48**, 153–162 (2015)
70. Ng, W.-S., Lee, M.-L., Hii, S.-L.: An overview of the factors affecting microbial-induced calcite precipitation and its potential application in soil im- provement. World Acad. Sci. Eng. Technol. **62**, 723–729 (2012)
71. Niemelä, S.I., Sivelä, C., Luoma, T., Tuovinen, O.H.: Maximum temperature limits for acidophilic, mesophilic bacteria in biological leaching systems. Appl. Environ. Microbiol. **60**(9), 3444–3446 (1994)
72. Nosouhian, F., Mostofinejad, D., Hasheminejad, H.: Influence of biodeposition treatment on concrete durability in a sulphate environment. Biosyst. Eng. **133**, 141–152 (2015). https://doi.org/10.1016/j.biosystemseng.2015.03.008
73. Ojha, A., Bandyopadhyay, T.K., Das, D., Dey, P.: Microbial carbonate mineralization: a comprehensive review of mechanisms, applications, and recent advancements. Mol. Biotechnol. (2025). https://doi.org/10.1007/s12033-025-01433-5
74. Okwadha, G.D.O., Li, J.: Optimum conditions for microbial carbonate precipitation. Chemosphere. **81**(9), 1143–1148 (2010). https://doi.org/10.1016/j.chemosphere.2010.09.066
75. Otlewska, A., Gutarowska, B.: Environmental parameters conditioning microbially induced mineralization under the experimental model conditions. Acta Biochim. Pol. **63**, 343–351 (2016)
76. Patwardhan, S.V., Tilburey, G.E., Perry, C.C.: Interactions of amines with silicon species in undersaturated solutions leads to dissolution and/or precipitation of silica. Langmuir. **27**(24), 15135–15145 (2011)
77. Park, H., Lee, K.Y.: Alginate/hyaluronate hydrogels for cartilage regeneration. J. Control. Release. **152**, e233–e234 (2011)
78. Patel, P.R., Patel, S.K.: Microbial concrete: the pioneering work on repairing concrete. Int. J. Pharm. Sci. Res. **2**(4), 825–828 (2011). https://doi.org/10.13040/IJPSR.0975-8232.2(4).825-28

79. Qiu, J., Tng, D.Q.S., Yang, E.-H.: Surface treatment of recycled concrete aggregates through microbial carbonate precipitation. Constr. Build. Mater. **57**(0), 144–150 (2014). https://doi.org/10.1016/j.conbuildmat.2014.01.085
80. Rahmaninezhad, S.A., Houshmand, M., Sadighi, A., Kamireddi, D., Street, R.M., Farnam, Y.A., Schauer, C.L., Raeisi_Najafi, A., Sales, C.M.: Generation of enhanced endospores for microbially induced calcium carbonate precipitation (MICCP) via thermal shock for concrete self-healing. Constr. Build. Mater. **419**(C) (2024)
81. Ramakrishnan V., Panchalan R.K., Bang S.S.: Bacterial Concrete. In: Proceedings of SPIE—International Society of Optical Engineering, vol. 4234, Australia (2001), pp. 168–176 (2001). https://doi.org/10.1117/12.4244404
82. Ramachandran, S.K., Ramakrishnan, V., Bang, S.S.: Remediation of concrete using microorganisms. ACI Mater. J. **98**(1) (2001)
83. Reddy, S.S.P., Rao, M.S., Aparna, P., Sasikala, C.: Performance of ordi-nary grade bacterial (*Bacillus subtilis*) concrete. Int. J. Earth Sci. Eng. **3**(1), 116–124 (2010)
84. Rodriguez-Navarro, C., Rodriguez-Gallego, M., Chekroun, K.B., Gonzalez-Munoz, M.T.: Conservation of ornamental stone by *Myxococcus xanthus*-induced carbonate biomineralization. Appl. Environ. Microbiol. **69**(4), 2182–2193 (2003)
85. Sahoo, K.K., Arakha, M., Sarkar, P., Jha, S.: Enhancement of properties of recycled coarse aggregate concrete using bacteria. Int. J. Smart Nano Mater. 1–17 (2016a)
86. Sahoo, K.K., Sathyan, A.K., Kumari, C., Sarkar, P., Davis, R.: Investigation of cement mortar incorporating *Bacillus sphaericus*. Int. J. Smart Nano Mater. 1–15 (2016b)
87. Sharma, B., Sharma, S., Reddy, M.S.: Ultrasonic monitoring of bacterial self-healing in concrete structures. In: Amziane, S., Toledo Filho, R.D., da Gloria, M.H.Y.R., Page, J. (eds.) *Bio-based building materials—proceedings of ICBBM 2025* Cham (2025)
88. Siddique, R., Chahal, N.: Effect of ureolytic bacteria on concrete properties. Constr. Build. Mater. **25**(10), 3791–3801 (2011). https://doi.org/10.1016/j.conbuildmat.2011.04.010
89. Siddique, R., Nanda, V., Kadri, E.-H., Khan, M.I., Singh, M., Rajor, A.: Influence of bacteria on compressive strength and permeation properties of concrete made with cement baghouse filter dust. Constr. Build. Mater. **106**, 461–469 (2016a). https://doi.org/10.1016/j.conbuildmat.2015.12.112
90. Siddique, R., Singh, K., Kunal, Singh, M., Corinaldesi, V., Rajor, A.: Properties of bacterial rice husk ash concrete. Constr. Build. Mater. **121**, 112–119 (2016b). https://doi.org/10.1016/j.conbuildmat.2016.05.146
91. Stocks-Fischer, S., Galinat, J.K., Bang, S.S.: Microbiological precipitation of $CaCO_3$. Soil Biol. Biochem. **31**(11), 1563–1571 (1999). https://doi.org/10.1016/s0038-0717(99)00082-6
92. Sriaporn, C., Komonjinda, S., Klayposri, N., Phinyo, K., Pekkoh, J., Awiphan, S., Chanprom, T.: Microbial and mineralogical characterization of the alkaline Chae Son hot spring. Northern Thailand. Extremophiles **29**(1), 2 (2024). https://doi.org/10.1007/s00792-024-01373-z
93. Skinner, H.C.W., Jahren, A.H.: Biomineralisation. In: Treatise on Geochemistry, vol. 8, pp. 117–184 (2003). https://doi.org/10.1016/B0-08-043751-6/08128-7
94. Sarayu, K., Iyer, N., Murthy, A.R.: Exploration on the biotechnological aspect of the Ureolytic bacteria for the production of the cementitious materials—a review. Appl. Biochem. Biotechnol. **172**(5), 2308–2323 (2014). https://doi.org/10.1007/s12010-013-0686-0
95. Sierra-Beltran, M.G., Jonkers, H.: Crack self-healing technology based on bacteria. J. Ceram. Process. Res. **16**(Special 1), 33–39 (2015)
96. Talaiekhozani, A., Keyvanfar, A., Shafaghat, A., Andalib, R., Majid, M.A., Fulazzaky, M.A., Zin, R.M., Lee, C.T., Hussin, M.W., Hamzah, N., Marwar, N.F., Haidar, H.I.: A review of self-healing concrete research development. J. Environ. Treat. Tech. **2**(1), 1–11 (2014)
97. Tiano, P., Cantisani, E., Sutherland, I., Paget, J.: Biomediated reinforcement of weathered calcareous stones. J. Cult. Herit. **7**(1), 49–55 (2006). https://doi.org/10.1016/j.culher.2005.10.003
98. Tobler, D.J., Cuthbert, M.O., Greswell, R.B., Riley, M.S., Renshaw, J.C., Handley-Sidhu, S., Phoenix, V.R.: Comparison of rates of ureolysis between *Sporosarcina pasteurii* and an indigenous groundwater community under conditions required to precipitate large volumes

of calcite. Geochim. Cosmochim. Acta. **75**(11), 3290–3301 (2011). https://doi.org/10.1016/j.gca.2011.03.023
99. Van Tittelboom, K., De Belie, N., De Muynck, W., Verstraete, W.: Use of bacteria to repair cracks in concrete. Cem. Concr. Res. **40**(1), 157–166 (2010). https://doi.org/10.1016/j.cemconres.2009.08.025
100. Van Wijngaarden, W., Vermolen, F., Van Meurs, G., Vuik, C.: Modelling biogrout: A new ground improvement method based on microbial-induced carbonate precipitation. Transp. Porous Media. **87**(2), 397–420 (2011)
101. Varenyam, A., Zhang, D.Y.: Bioremediation of Pb-contaminated soil based on microbially induced calcite precipitation. J. Microbiol. Biotechnol. **22**(2), 244–247 (2012)
102. Veluchamy, C., Sharma, A., Thiagarajan, K.: Assessing the impact of heavy metals on bacterial diversity in coastal regions of southeastern India. Environ. Monit. Assess. **196**(9), 828 (2024)
103. Wang, J., De Belie, N., Verstraete, W.: Diatomaceous earth as a protective vehicle for bacteria applied for self-healing concrete. J. Ind. Microbiol. Biotechnol. **39**(4), 567–577 (2012). https://doi.org/10.1007/s10295-011-1037-1
104. Wang, J., Soens, H., Verstraete, W., De Belie, N.: Self-healing concrete by use of microencapsulated bacterial spores. Cem. Concr. Res. **56**, 139–152 (2014). https://doi.org/10.1016/j.cemconres.2013.11.009
105. Weiner, S., Dove, P.M.: An overview of biomineralization processes and the problem of the vital effect. Rev. Mineral. Geochem. **54**(1), 1–29 (2003). https://doi.org/10.2113/0540001
106. Whiffin, V.S.: Microbial $CaCO_3$ Precipitation for the Production of Biocement. Unpublished doctoral dissertation, Murdoch University (2004)
107. Wong, L.S.: Microbial cementation of ureolytic bacteria from the genus bacillus: A review of the bacterial application on cement-based materials for cleaner production. J. Clean. Prod. **93**, 5–17 (2015). https://doi.org/10.1016/j.jclepro.2015.01.019
108. Zeikus, J.: Thermophilic bacteria: ecology, physiology and technology. Enzym. Microb. Technol. **1**(4), 243–252 (1979). https://doi.org/10.1016/0141-0229(79)90043-7
109. Zeigler, D.R.: The Genus Geobacillus Introduction and strain catalog (7th ed., Vol. Volume 3: The Genus Geobacillus). The Bacillus Genetic Stock Centre, Department of Biochemistry, The Ohio State University, 484 West Twelfth Avenue, Columbus, Ohio 43210: Bacillus Genetic Stock Centre (2001)
110. Zhang, R., Ge, Y., Wang, C., Etim, I.-I.N., Khan, S., Li, C., Yang, L., Liu, J., Liu, J., Yi, P.: Deep-sea microbially influenced corrosion and biomineralization. Frontiers in microbiology **16**, 1605909 (2025)
111. Zhang, Y., Li, B., Liu, J., Han, D., Rohani, S., Gao, Z., Gong, J.: Inhibition of crystal nucleation and growth: A review. Cryst. Growth Des. **24**(6), 2645–2665 (2024). https://doi.org/10.1021/acs.cgd.3c01345
112. Zhang, J., Wu, R., Li, Y., Zhong, J., Deng, X., Liu, B., Han, N.X., Xing, F.: Screening of bacteria for self-healing of concrete cracks and optimization of the microbial calcium precipitation process. Appl. Microbiol. Biotechnol. 1–10 (2016)
113. Zhou, T., Dong, Z., Setlow, P., Li, Y.: Kinetics of germination of individual spores of *Geobacillus stearothermophilus* as measured by Raman spectroscopy and differential interference contrast microscopy. PLoS One. **8**(9) (2013)

# Chapter 3
# Application of Self-Healing Agents in Concrete and Its Properties

## 3.1 Introduction

Self-healing allowing concrete and mortar to retain its serviceability over extended periods. This chapter begins by exploring three (3) methods of applying self-healing agents for facilitating improvement in permeability and filling cracks for new and existing concrete by microbially induced calcite precipitation (MICP). The self-healing methods include direct incorporation, surface treatment, and encapsulation of healing agents prior to their addition into concrete or mortar. In the direct incorporation and encapsulation approaches, the self-healing agents are introduced into the concrete mix during the mixing process. In contrast, the surface treatment method involves the deposition of bacterial calcite on the existing concrete surface to fill cracks and prevent the ingress of harmful substances, thereby mitigating further deterioration. This chapter examines the effects of these methods on the physical, mechanical, and durability properties of concrete, as reported by previous researchers and works performed by the authors. It also provides insights into how the integration of self-healing agents can enhance the overall performance of concrete and mortar.

The use of microbially induced calcite precipitation (MICP) as microbial concrete has become increasingly popular as it is natural, environmentally friendly and economical in the long run. The effect of MICP on the cementitious material has proven in several experimental works. The studies grouped in three (3) application categories in which the way of bacteria introduced to the concrete, namely direct incorporation, surface treatment and by encapsulation method. A review on each category are outlined in the following sections.

 
R. M. A. M. Azam and D. H. M. Saman, *Self-Healing Concrete*, Engineering Materials,
https://doi.org/10.1007/978-981-95-7553-4_3

## 3.2 Direct Incorporation of Bacteria into Concrete

Direct incorporation of bacteria into concrete defined as the direct addition of bacteria into concrete mix during the mixing process. The suitable types of bacteria to be incorporated should have the ability to endure a high alkaline environment in concrete due to high pH and able to produce endospore with the mechanical stress during the mixing process [27, 49]. This method was first introduced by Gollapudi et al. [33] as a means of preventing leaching in channels through crack restitution. In their study, *Sporosarcina pasteurii* bacterial strains were used to plug a sand column. The results showed that higher concentrations of bacteria led to a 100% reduction in permeability, whereas lower concentrations were less effective. The presence of bacteria significantly enhanced calcite deposition, while no calcite formation was observed in the absence of bacteria. Similar bacteria also used in a study conducted by Ramachandran et al. [74] with different concentrations of bacteria ranging from $3.8 \times 10^3$ to $7.6 \times 10^6$ cfu/ml. An increase in concrete strength was observed, ranging from 11 to 16% between 7 and 28 days of curing. However, a decline in strength was noted across all concentrations beyond 28 days. A similar decreasing trend after 28 days was also reported by Park et al. [68], who used *Sporosarcina pasteurii* in their study. The results showed a 17.3% reduction in strength compared to the strength at seven (7) days of age. This decline may be attributed to changes in biomass within the mortar matrix, as the direct incorporation of bacteria typically results in a shorter lifespan and a significant reduction in calcite production over time. In a related study, Achal et al. [2] investigated the incorporation of *Bacillus sphaericus* in concrete. The addition of this microbial agent increased the compressive strength of mortar by 36%. Furthermore, due to microbial calcite deposition, the mortar with microbes exhibited six (6) times lower water absorption compared to the control specimens.

In a related experiment, Jonkers and Schlangen [42] also investigated the incorporation of *Bacillus pseudofirmus* into the mortar matrix. The mortar specimens were incubated in a nutrient medium containing yeast extract and peptone. Environmental Scanning Electron Microscope (ESEM) analysis revealed that the bacterial cement matrix precipitated a substantial amount of calcite-like crystals on its surface when incubated in the nutrient medium, compared to the control specimens. This study aimed to evaluate the bacteria's ability to precipitate calcite using nutrients applied externally as part of the incubation medium. The authors concluded that the ability of bacteria to metabolise external nutrients suggests that such nutrients should ideally be integrated into the concrete matrix to enhance calcite precipitation.

In continuation of the earlier study by Jonkers and Schlangen [42], *Bacillus pasteurii* at a concentration of $10^9$ cfu/ml was incorporated into concrete in a subsequent study by the same authors [43]. In this study, mortar containing *Sporosarcina pasteurii* was compared with mortar mixes supplemented with various nutrients intended to support bacterial growth, including sodium aspartate (Na-aspartate), sodium glutamate (Na-glutamate), sodium polyacrylate (Na-polyacrylate), and sodium citrate (Na-citrate). The aim was to identify the most suitable nutrient that would enhance bacterial activity without compromising the durability of the concrete.

The results indicated no significant difference in compressive strength among the control mortar, the mortar containing *Sporosarcina pasteurii*, and those supplemented with Na-aspartate and Na-glutamate. However, mortars incorporating Na-polyacrylate and Na-citrate exhibited a significant reduction in compressive strength. These findings suggest that the selection of appropriate bacterial strains along with compatible nutrients is crucial to maintaining the durability and structural integrity of concrete.

Ghosh et al. [32] incorporated *Shewanella* that was isolated from crude oil samples from a hot spring at Bakreshwar, West Bengal, India into the mortar. The incorporation of Shewanella into mortar were conducted on concentration of $10^2$, $10^3$, $10^4$, $10^5$, $10^6$ and $10^7$ cfu/ml, respectively. Based on experiments conducted, the increment of compressive strength maximised at the bacterial cell concentration of $10^5$ cfu/ml. The bacterial concentration of more than $10^5$ cfu/ml, which were $10^6$ and $10^7$ cfu/ml decreased the compressive strength result at 8.3% and 19.4%, respectively at 28 days of age. Such increment only found at a particular concentration of bacteria and compounded with the higher incorporation of bacteria. Such condition also attained in a study by Eisa et al. [28], Sahoo et al. [77], Reddy et al. [76], Ghosh et al. [31], Ghosh et al. [30], and Ramachandran et al. [74]. This phenomenon might attribute to Growth Advantage in Stationary Phase (GASP) [38, 52, 108]. GASP defined by Bačun-Družina et al. [10] as a condition of non-sporulating bacteria respond to the starvation conditions by undergoing a series of morphological and physiological changes. Firstly, GASP resulted bacteria cells failing to adapt to new circumstances, that in turn, a cell population enters death phase, and second, a cell population evolves mechanisms enabling it to rapidly adapt to stress conditions and remain viable during prolonged stationary growth phase. This competitive state would influence the calcite precipitation that contributes to the compressive strength enhancement. The higher the bacteria concentration, GASP phenotype condition would also be increased. Therefore, it reflects why only the particular concentration level of bacteria would maximise the compressive strength results.

Bacteria have also been incorporated into fly ash-based concrete. The first study on this application was conducted by Achal et al. [3], which aimed to evaluate the effects of adding *Bacillus megaterium* at a concentration of $5 \times 10^7$ cfu/ml on the compressive strength and water absorption of fly ash mortar. Fly ash was used to replace 10, 20, and 40% of the total cement weight in the mortar mixes. The results showed improvements in compressive strength of 19%, 14%, and 10%, respectively, compared to the control specimens. Additionally, bacterial incorporation resulted in a threefold reduction in water absorption, attributed to microbial calcite precipitation that densified the mortar matrix. Similarly, in a study by Chahal et al. [15], the inclusion of *Sporosarcina pasteurii* at a concentration of $10^5$ cfu/ml led to a 22% increase in compressive strength compared to control specimens. Moreover, due to microbial calcite deposition, water absorption in the bacterial mortar was more than four (4) times lower than in the control samples.

A study by Sahoo et al. [77] has conducted on the evaluation of concrete performance using recycled coarse aggregate (RCA) with an addition of the *Bacillus subtilis* regarding compressive strength, capillary water absorption and drying shrinkage.

The compressive strength of RCA concrete has increased by 20% with the $10^6$ cfu/ml incorporation of *Bacillus subtilis*. Moreover, the inclusion of bacteria was also causing the reduction in capillary water absorption and drying shrinkage. In contrast, direct incorporation of bacteria not only was found to be effective in improving concrete durability but also in remediation of cracks. According to Krishnapriya et al. [49], the integration of *Bacillus megaterium*, *Bacillus licheniformis* and *Bacillus flexus* at $10^6$ cfu/ml in concrete has positively improved the strength at 12.1%, 10.6% and 6.1% respectively, when compared to control specimens. On the other side, concrete incorporated with *Bacillus megaterium* and *Bacillus licheniformis* have completed healing of cracks after 81 days. However, only partial healing was observed in concrete that combined with *Bacillus flexus*. *Bacillus megaterium* was also incorporated directly in concrete by Andalib et al. [7] using different bacterial concentrations ($10 \times 10^5$ to $50 \times 10^5$ cfu/ml). The results revealed that the bacterial concentration of $30 \times 10^5$ cfu/ml showed a maximum increment in the compressive and flexural strength of structural concrete. The strength of the highest grade of structural bacterial concrete improved as compared to lower grade due to the higher amount of calcite precipitation. The maximum rate of strength development (24%) achieved in the highest grade of structural concrete (50 MPa) due to the maximum rate of $CaCO_3$ precipitation intensity.

Some of the data on concrete durability showed that positive result attributed until 28 days of age and some of the data showed longer than 28 days. The bacteria types, bacterial concentrations and incorporation method would determine the survivability of bacteria. As described by Ramachandran et al. [74], during the initial curing period, the mortar was still porous, which enable the live cells obtained real nourishment. With right food, the calcite precipitation resulted from cell growth occurred on the cell surface and within the cement matrix. Hence, resulting in less porous and permeable mortar which limiting the bacteria to obtain the nourishment. The choice of microorganism plays a significant role in the enhancement of concrete durability and performance. The comparison between compressive strength and flexural strength along with improvement made of mortar/concrete by direct incorporation of bacteria at 28 days are summarised in Table 3.1 and Table 3.2 respectively.

A significant increase concerning compressive strength with the presence of bacteria has been observed [1, 42–44]. Most of the bacteria with genus *Bacillus* is aerobic (bacteria growth in the presence of oxygen, $O_2$). The denser mortar was resulted by continuous hydration process along with calcite which plugged pores. The dense mortar inhibited the oxygen penetration that acquired for bacterial activity. Therefore, the bacterial cells would die or turn into endospore. High alkaline in the mortar matrix, especially during hydration where pH greater than 12 also contribute to the reduction of bacterial cells [12]. Cracks formed after the mortar has densified can reduce the viability of bacterial cells. All of these justifies why limited data on the direct incorporation of bacteria into concrete found for crack remediation. Most studies have used high bacterial concentrations ($10^5$ cfu/mL and above), as lower cell counts are generally insufficient to effectively demonstrate the potential of bacteria in crack remediation.

**Table 3.1** Compressive strength of concrete at age 28 days and improvement made by direct incorporation of bacteria reported by previous researchers

| References | Bacteria | Cell concentration (cfu/ml) | Compressive strength at 28 days (MPa) | % of increase with reference to control specimens |
|---|---|---|---|---|
| Andalib et al. [7] | *Bacillus megaterium* | Control | 33.0* | – |
| | | $10 \times 10^5$ | 35.0* | 5.7 |
| | | $20 \times 10^5$ | 36.0* | 8.3 |
| | | $30 \times 10^5$ | 38.0* | 13.2 |
| | | $40 \times 10^5$ | 37.0* | 10.8 |
| | | $50 \times 10^5$ | 34.0* | 2.9 |
| Sahoo et al. [77] | *Bacillus subtilis* | Control | 38.2 | – |
| | | $1 \times 10^1$ | 41.0 | 6.8 |
| | | $1 \times 10^3$ | 43.1 | 11.4 |
| | | $1 \times 10^6$ | 46.2 | 17.3 |
| | | $1 \times 10^7$ | 44.6 | 14.3 |
| Sahoo et al. [78] | *Bacillus sphaericus* | Control | 5.9 | – |
| | | $1 \times 10^5$ | 6.9 | 18.3 |
| | | $1 \times 10^6$ | 7.0 | 18.9 |
| | | $1 \times 10^7$ | 7.3 | 23.4 |
| | | $1 \times 10^8$ | 6.2 | 4.9 |
| | | $1 \times 10^9$ | 6.1 | 3.4 |
| Krishnapriya et al. [49] | | Control | 33.0 | – |
| | *Bacillus megaterium* | $1 \times 10^5$ | 37.0 | 10.8 |
| | *Bacillus licheniformis* | $1 \times 10^5$ | 36.5 | 9.6 |
| | Bacillus *flexus* | $1 \times 10^5$ | 35.0 | 5.7 |

(continued)

**Table 3.1** (continued)

| References | Bacteria | Cell concentration (cfu/ml) | Compressive strength at 28 days (MPa) | % of increase with reference to control specimens |
|---|---|---|---|---|
| Chahal et al. [15] | *Sporosarcina pasteurii* | Control | 24.0 | – |
| | | $1 \times 10^3$ | 25.0 | 4.2 |
| | | $1 \times 10^5$ | 28.0 | 16.7 |
| | | $1 \times 10^7$ | 26.0 | 8.3 |
| Achal et al. [3] | *Bacillus megaterium* | Control | 23.0* | – |
| | | $5 \times 10^7$ | 28.0* | 17.9 |
| Jonkers et al. [45] | *Bacillus pseudofirmus* | Control | 51.0* | – |
| | | $6 \times 10^8$ | 55.0* | 7.3 |
| Jonkers and Schlangen [43] | *Sporosarcina pasteurii* | Control | 31.9 | – |
| | | $1 \times 10^9$ | 34.8 | 9.0 |

* Interpolated and derived from graph

**Table 3.2** Flexural strength of concrete at 28 days of age and improvement made by direct incorporation of bacteria reported by previous researchers

| References | Bacteria | Cell concentration (cfu/ml) | Compressive strength at 28 days (MPa) | % of increase with reference to control specimens |
|---|---|---|---|---|
| Andalib et al. [7] | *Bacillus megaterium* | Control | 3.7* | – |
| | | $10 \times 10^5$ | 3.9* | 5.1 |
| | | $20 \times 10^5$ | 4.0* | 7.9 |
| | | $30 \times 10^5$ | 4.1* | 9.8 |
| | | $40 \times 10^5$ | 3.9* | 5.1 |
| | | $50 \times 10^5$ | 3.8* | 2.6 |
| Ravindranath and Likhit [75] | *Sporosarcina pasteurii* | Control | 7.1 | – |
| | | ND | 7.9 | 10.1 |
| Jonkers and Schlangen [42] | *Sporosarcina pasteurii* | Control | 7.78 | – |
| | | $1 \times 10^9$ | 7.45 | −4.4 |

ND: Not described
* Interpolated and derived from graph

The effect of direct bacterial incorporation on water absorption at optimum cell concentrations, along with the percentage improvement compared to control specimens, is presented in Table 3.3. Siddique et al. [81] used *Bacillus aerius* and reported a 33.3% reduction in water absorption. In a separate study by Siddique et al. [82], also using *Bacillus aerius*, an even greater reduction of 50% in water absorption was recorded. On the other hand, Nosouhian et al., [66] compared the performance of three (3) types of bacteria namely *Bacillus sphaericus*, *Bacillus subtilis* and *Sporocarsina pasteurii* towards water absorption. The results obtained portrayed that only *Bacillus sphaericus* reduced water absorption by 1.5% while those with the other two bacteria increased the water absorption rate at 1.4%.

In the following study, the authors incorporated *Geobacillus stearothermophilus* directly into the mortar, and the results are reported accordingly. Three (3) testing were performed to evaluate the effect of incorporating *Geobacillus stearothermophilus* (D-GS) directly into mortar namely compressive strength, flexural strength and water absorption. The compressive strength test was performed to evaluate the imprint of *Geobacillus stearothermophilus* directly incorporated inside the cement-sand mortar (D-GS). Six (6) concentrations of bacteria were tested and compared with the control mortar. Table 3.4 depicts the relative compressive strength results for D-GS mortar corresponding to control mortar, respectively. Figure 3.1 shows the trend of compressive strength for D-GS mortar specimens versus age.

From the compressive strength results given in Table 3.4 and Fig. 3.1, mortar incorporated with *Geobacillus stearothermophilus* performed better than that of control

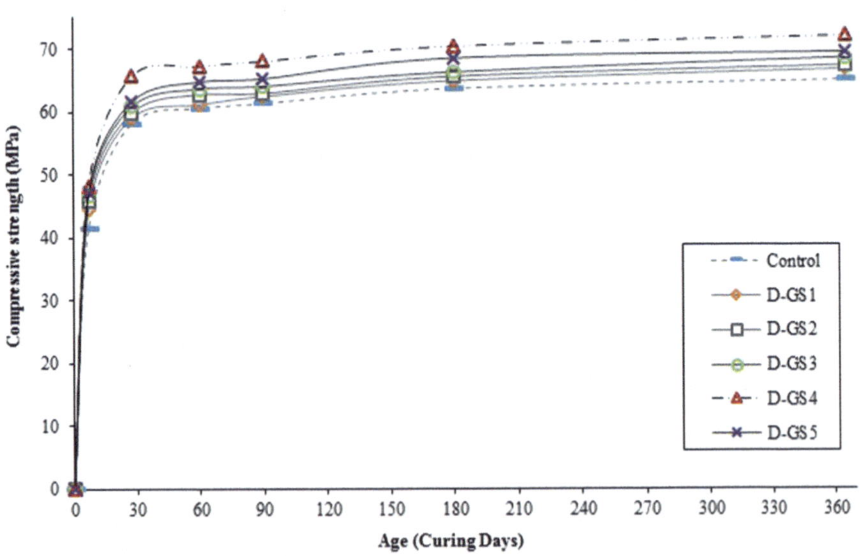


**Fig. 3.1** Compressive strength of six (6) series of D-GS mortar containing different cell concentrations of *Geobacillus stearothermophilus* versus age

**Table 3.3** Rate of water absorption and improvement made to mortar/concrete at 28 days by direct incorporation of bacteria reported by previous researchers

| References | Bacteria | Cell concentration (cfu/ml) | Compressive strength at 28 days (MPa) | % of decrease with reference to control specimens |
|---|---|---|---|---|
| Siddique et al. [81] | *Bacillus aerius* | Control | 1.56* | – |
| | | 1 x $10^5$ | 1.2* | −33.3 |
| Siddique et al. [82] | *Bacillus aerius* | Control | 2.1* | – |
| | | 1 x $10^5$ | 1.4* | −50.0 |
| Nosouhian et al. [66] | | Control | 6.8 | – |
| | *Bacillus sphaericus* | ND | 6.7 | −1.5 |
| | *Bacillus subtilis* | ND | 6.9 | 1.4 |
| | *Sporocarsina pasteurii* | ND | 6.9 | 1.4 |

ND: Not described

mortar. Seemingly, the mortar specimens designated D-GS4 which contained bacterial concentration of 1 × $10^9$ cfu/ml attained highest compressive strength to that of control mortar for about 16.4%, 13.1%, 11.2%, 11.1%, 10.7% and 10.9% corresponding to those specimens at 7, 28, 60, 90, 180 and 365 days of age respectively. Higher than cell concentration of 1 × $10^9$ cfu/ml, the trend reversed.

Two factors contribute to the observed opposite trend. Firstly, a high concentration of bacteria can create excessive voids within the concrete matrix, and the resulting formation of calcium carbonate ($CaCO_3$), while beneficial for strength development, may not be sufficient to compensate for these additional gaps. Secondly, the reduction in strength at higher bacterial concentrations is attributed to the bacteria's adaptation to low-nutrient environments, known as the Growth Advantage in Stationary Phase (GASP) phenotype [38, 108]. As the number of bacteria populations increases, the restriction on nutrients also increases. These bacteria will generate a starvation signal which in turn they changed into stationary phase. GASP indicates the stationary phase as a starvation response by bacteria, the bacteria will remain homogeneous, in which some of the bacteria may turn into death phase, and some bacteria remain viable [108]. Furthermore, it also attributed to the behaviour of biomass within the mortar matrix. Upon cell growth, the precipitation of calcite would occur on the cell surfaces and the mortar matrix as well. Hence, the mortar became less porous and permeable in which the calcite has plugged it. Eventually, this will retard the flow of nutrients and oxygen to the bacterial cells; which the bacteria would be died or turned into endospores. This explanation defines on the reduction of mortar properties enhancement with the incorporation of *Geobacillus stearothermophilus* of more than 1 × $10^9$ cfu/ml.

The improvement of compressive strength is attributed to deposition of precipitation from bacteria within the cement matrix. The deposition from bacteria has plugged the smaller pores and densified in the cement matrix thus increase the compressive strength of D-GS mortar batch as compared to control. The significant increase marked at the age of seven (7) days. However, a marginal increase in compressive strength with the incorporation of *Geobacillus stearothermophilus* into mortar for subsequent curing days for all concentrations of bacteria might be ascribed to the viability of cells at an early age of curing. Direct incorporation of bacteria diminished the viability of bacteria as further hydration followed by reduction of pore sizes

**Table 3.4** Compressive strength for D-GS mortar specimens containing different cell of *Geobacillus stearothermophilus* with relative to the control mortar

| Exp. No. | Batch | BC (cfu/ml) | % relative with respect to control | | | | | |
|---|---|---|---|---|---|---|---|---|
| | | | 7 days | 28 days | 60 days | 90 days | 180 days | 365 days |
| 1 | Control | – | – | – | – | – | – | – |
| 2 | D-GS1 | 1 × $10^3$ | 7.2 | 1.5 | 1.0 | 1.8 | 1.9 | 2.6 |
| 3 | D-GS2 | 1 × $10^5$ | 10.4 | 2.9 | 3.6 | 2.6 | 3.0 | 3.5 |
| 4 | D-GS3 | 1 × $10^7$ | 12.5 | 4.6 | 5.1 | 4.4 | 4.1 | 5.4 |
| 5 | D-GS4 | 1 × $10^9$ | 16.4 | 13.1 | 11.2 | 11.1 | 10.7 | 10.9 |
| 6 | D-GS5 | 1 × $10^{11}$ | 13.7 | 6.0 | 6.9 | 6.4 | 7.5 | 6.8 |

as reported by Bang et al. [11] that in turn increased the dead cells in the concrete matrix. Nevertheless, the organic matter along with biomass of the dead cells of bacteria would become as a natural fibre within the concrete matrix [15, 74] that bridging the pores, thus, increase the strength development.

The performance of mortar with direct incorporation of *Geobacillus stearothermophilus* is within the range of improvement which are 13.1% with $1 \times 10^9$ cfu/ml as reported earlier. Therefore, an increment of compressive strength of mortar with the incorporation of *Geobacillus stearothermophilus* in enhancing mortar strength were comparable with those made of other types of bacteria that reported previously.

Five (5) series of mortar with direct incorporation of *Geobacillus stearothermophilus* (D-GS) also undergone flexural strength test which comprises of 108 mortar specimens tested. Six (6) concentrations of bacteria were tested including control mortar. Table 3.5 depicts the flexural strength for mortar with direct incorporation of *Geobacillus stearothermophilus* (D-GS) and comparison of flexural strength and improvement made by other types of bacteria at 28 days.

Figure 3.2 shows the trend of flexural strength test results with age. Based on Table 3.5 and Fig. 3.2, the incorporation of *Geobacillus stearothermophilus* in mortar increased the flexural strength. The flexural strength increase for D-GS4 mortar with reference to those of control mortar are 16.8%, 19.9%, 11.9%, 13.1%, 10.6% and 7.5% at 7, 28, 60, 90, 180 and 365 days of age respectively, the highest increment recorded among the concentration of bacteria. The highest strength obtained by batch D-GS4 mortar ($1 \times 10^9$ cfu/ml). The mortar that contained a bacterial concentration more than $1 \times 10^9$ cfu/ml implies the GASP phenotype. GASP restricts the growth of bacteria by limitation on nutrient available; thus, lower the flexural strength of the resulted mortar.

Higher flexural strength was obtained by inclusion of bacteria at $30 \times 10^5$ cfu/ml concentration with an improvement of 9.8% to the control one [7]. It was comparable with those of *Geobacillus stearothermophilus* incorporated at $1 \times 10^5$ cfu/ml recorded improvement by 9.4% to the control one. Research conducted by Ravindranath and Likhit [75] has proven that incorporation of bacteria improved the flexural strength slightly as compared to control mortar. Mortar incorporated with *Sporocarsina pasteurii* directly attained a strength of 7.9 N/mm$^2$ as compared to control

**Table 3.5** Flexural strength for D-GS mortar specimens containing different cells of *Geobacillus stearothermophilus* with relative to the control mortar

| Exp. No. | Batch | BC (cfu/ml) | % relative with respect to control | | | | | |
|---|---|---|---|---|---|---|---|---|
| | | | 7 days | 28 days | 60 days | 90 days | 180 days | 365 days |
| 1 | Control | – | – | – | – | – | – | – |
| 2 | D-GS1 | $1 \times 10^3$ | 2.7 | 6.1 | 3.2 | 1.7 | 1.7 | 1.2 |
| 3 | D-GS2 | $1 \times 10^5$ | 6.7 | 9.4 | 4.3 | 5.5 | 4.6 | 2.2 |
| 4 | D-GS3 | $1 \times 10^7$ | 8.1 | 11.6 | 8.7 | 7.2 | 5.6 | 2.8 |
| 5 | D-GS4 | $1 \times 10^9$ | 16.8 | 19.9 | 11.9 | 13.1 | 10.6 | 7.5 |
| 6 | D-GS5 | $1 \times 10^{11}$ | 13.4 | 15.5 | 9.9 | 7.9 | 6.9 | 4.7 |

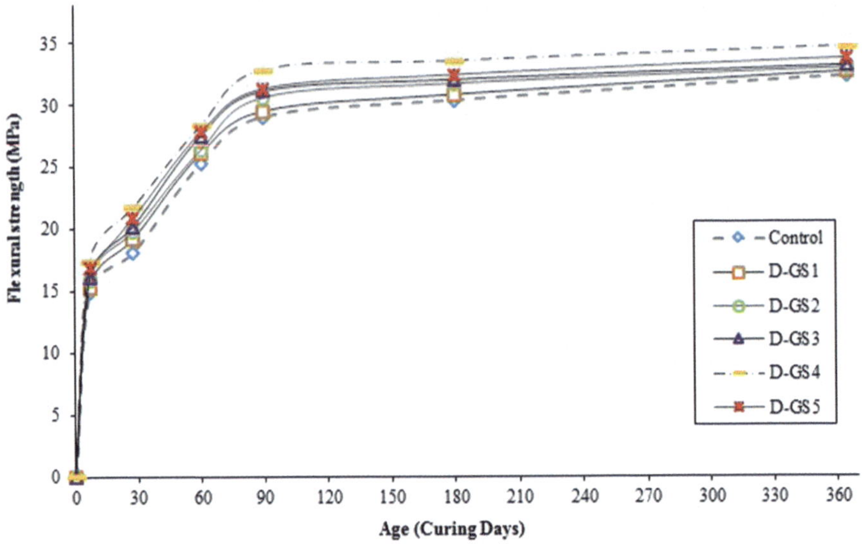


**Fig. 3.2** Flexural strength of mortar specimens containing direct incorporation of different concentration of *Geobacillus stearothermophilus* versus age

mortar (without bacteria) charted 7.1 N/mm$^2$ of flexural strength. However, contradicted result was reported by Jonkers and Schlangen [42], incorporating *Sporocarsina pasteurii* at 1 × 10$^9$ cfu/ml into mortar, found slightly decrease in flexural strength by 4.4% as compared to those of control specimen.

Water absorption test was also performed on mortar specimens that directly incorporated with different concentrations of bacteria *Geobacillus stearothermophilus* (D-GS) to measure the water absorption rate as compared to control mortar. Figure 3.3 demonstrates the trend of water absorption results tested on 108 specimens of mortar. The incorporation of *Geobacillus stearothermophilus* found to decrease the water absorption capacity of the produced mortar. Those with bacterial concentration 1 × 10$^9$ cfu/ml namely D-GS4 depicts the lowest absorption marked 22.4%, 23.3%, 22.6%, 31.7%, 41.4% and 63.2% lower than that of control mortar at 7, 28, 60, 90, 180 and 365 days, respectively. The D-GS4 batch was also observed having the highest compressive strength and flexural strength. It was confirmed that the presence of *Geobacillus stearothermophilus* able to decrease water absorption rate of the mortar. The ability of bacteria to produce calcium carbonate has filled some percentage of void volume thereby making the matrices more compact and resistive to seepage. Besides, a significant impact on the development of $CaCO_3$ crystal morphology influenced by the bacterial types and medium composition used. De Muynck et al. [22, 23] emphasised on a notable decrease in uptake of water, with a less pronounced change in the chromatic aspect resulted by using pure cultures. A pure culture of *Geobacillus stearothermophilus* used through-out the mixing process;

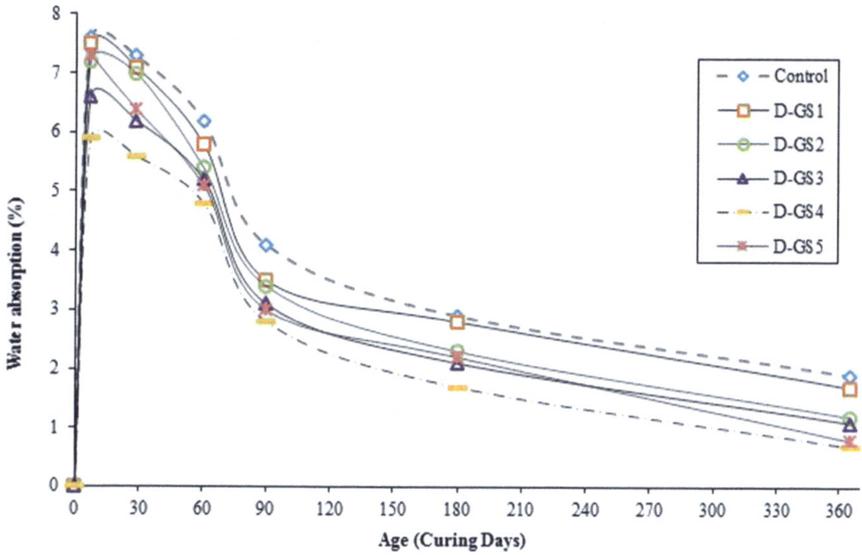


**Fig. 3.3** Water absorption for mortar specimens containing direct incorporation of different concentration of *Geobacillus stearothermophilus*

which in turn improve the water uptake as compared to non-bacterial mortar specimens. Research conducted by Wang et al. [96], Chahal et al. [15], De Myunck et al. [22, 23] concluded that the inclusion of bacteria in the concrete mix has able to increase the performance of concrete towards pore deposition of calcium carbonate crystals in which able to decrease the water absorption. However, direct incorporation method only applies for new concrete. The following sub-section describes how surface treatment using a bacterial solution can be applied to fill cracks in existing concrete or mortar.

## 3.3 Surface Treatment

Good quality of the concrete surface is not only visually pleasing, but more importantly, it serves as a barrier to protect any deleterious material from seep through inside concrete to avoid further deterioration. Chemicals such as silane [5, 19, 40, 56, 63], siloxane [5, 60, 88], silicate [55, 64] and acrylate [5, 102] were among of the concrete surface treatment that have been used conventionally. However, these types of chemicals pose a serious harm to the environment as it considered as toxic substances. Therefore, bacterial calcite deposition was attempted as one of the methods for surface crack treatment [54, 74].

Ramachandran et al. [74] conducted a study on various depths of cracks along with the stiffness values to evaluate the efficiency of healing by bacteria. Based on observation, the presence of *Sporocarsina pasteurii* increasing the stiffness of beams by 9.4% with a 3.175 mm cut and by 4.8% in 9.525 mm cut as compared to the controls. It showed that the remediation in the shallower crack was more efficient rather than in deeper ones. In conjunction with Ramachandran et al. [74] work, several studies conducted on surface treatment using bacterial deposition carried out and discussed as follows.

De Myunck et al. [22] used pure *Bacillus sphaericus*, mixed ureolytic bacterial culture solution and conventional treatment (silanes and siloxanes) to treat the mortar surface crack. Different treatments were applied to gain a better insight on bacterial treatment as compared to conventional. Based on the observation, the deposition of a layer of calcite due to bacterial precipitation on the specimen surface has resulted in a decrease of capillary suction and a reduction in gas permeability. The use of pure cultures also results in a more pronounced reduction in the uptake of water, with a less noticeable change in the chromatic aspect when compared to the use of mixed ureolytic cultures as a paste. The overall results which concerning capillary water uptake, gas permeability, the chromatic element has obtained with pure cultures of the *Bacillus sphaericus* species were comparable to the ones achieved with conventional water repellents (silanes, siloxanes). On the other hand, *Sporocarsina pasture* also were studied by Qiu et al. [72] as a surface treatment of concrete amended with recycled concrete aggregates (RCA) through microbial carbonate precipitation (MCP). Several conclusions made in the study, which is the type of bacteria used can precipitate $CaCO_3$ crystals on RCA and reduce its water absorption when $CaCl_2$ and nutrient is provided. The amount of MCP peaks of $pH = 9.5$ and increased with higher temperature, bacteria concentration, or calcium concentration. The rate, amount, and type of MCP on RCA concrete able to be modified through proper control of culturing and precipitating conditions. The author reported no significant depth of cracks treated by using this method. Another study conducted by Nosouhian et al. [66], using *Bacillus sphaericus* at $6.6 \times 10^6$ cfu/ml, *Bacillus subtilis* at $4.9 \times 10^5$ cfu/ml and *Sporocarsina pasteurii* at $9.7 \times 10^7$ cfu/ml were treated on mortar crack which later being exposed to sulphate environment. He discovered that the expansion rate of the bacteria treated specimens with *Bacillus pasteurii* exposed to sulphate environments reduces concrete degradation the most. Besides, due to the lower surface permeability attributed to the existence of a calcium carbonate layer on the specimen, the average water absorption of bacterial treated groups decreased up to 12.14% over control groups at 210 days of age. A slight increase of the concrete compressive strength observed with the bacterial surface treatment with no surface treated depth was recorded.

Sangadji et al. [79] used alkaliphilic strain species from the genus *Bacillus* solution for crack remediation. The liquid bacteria-based solution was injected into the porous network until it reached and flew out through the crack opening. The bacteria then precipitated calcium carbonate to block the crack. Permeability reduction up to 90% has achieved at 11 days incubation period. However, the permeability reduction rate drops to 70% at 19 days of incubation due to incomplete flushing of the solution throughout the crack opening. The results showed that wet cured bacteria series exhibited 99% permeability reduction as compared to dry cured specimen charted only 60% reduction in permeability. Wiktor and Jonkers [99] also used alkaliphilic species from the genus *Bacillus* as bacteria based repair system solution. The solution sprayed on the surface of cracks and a concrete slab in an underground car park. Water permeability and freeze/thaw resistance tests were performed to assess the crack-sealing efficiency and improvement of frost salt scaling. Only cracks that had not treated with the bacteria-based repair system were still heavily leaking. Also, the freeze/thaw resistance of concrete that treated with the bio-based repair system was significantly higher than the untreated concrete.

On the other hand, Qian et al. [73] used *Bacillus mucilaginous* L3 with a concentration of $2.3 \times 10^9$ cfu/ml on the restoration of micro-pores and cracks. Cracks brushed with an agar solution prior spraying the bacteria over the agar solution to the crack specimen surface. This method ensures the bacteria immobilised in the agar layer onto the sample surface. The nutrient solutions for bacteria, namely calcium hydroxide, ($Ca(OH)_2$) and sodium bicarbonate, ($NaHCO_3$) sprayed on the cement specimens every three (3) days. Control samples were prepared using the same concentration of bacteria with the same nutrition sprayed onto the surface, but without agar to immobilise the bacteria. The results obtained revealed that at the crack width of 50 μm to 100 μm, a dense and continuous deposited layer was formed on the surface specimens which bonded the cement matrix tightly. However, un-immobilised bacteria that applied onto the crack surface witness lower quality of bond strength between deposited layer and specimen surface that can quickly spall by minimum scratching. The application of an agar layer to immobilise bacteria has created a suitable environment so that the bacteria able to perform biomineralisation process. The combination of agar and mineralised particle has proven to form a dense film as compared to the presence of mineralised particle independently.

Bacteria based treatment to heal surface crack with several types of bacteria under genus *Bacillus* has been reported. The surface treatment applied to concrete surface decrease the crack width, improved the macropores resulting increase resistance of deleterious material from ingress to the concrete [54, 66, 73, 79]. However, the limitation of this type of healing treatment is that it will be only applicable on the observed and visible cracks, hence the maintenance process still needs to be undertaken by monitoring the occurrence of cracks. The depth penetration of each spray treatment should also be considered in the future as no report in this matter has obtained in the literature. The surface crack closure does not confirm the same closing effect towards the interior concrete structure.

More investigation works on crack closures using bacteria broth solution by surface treatment are elaborated in Chap. 4. The bacteria could perform better when protected, in which the elimination of external factor that would inhibit its growth. Due to this limitation, the researchers have looked for a further realistic approach by having the healing process from inside to outside of concrete or mortar. The elaboration of such healing method is discussed in the following sub-sections.

## 3.4 Encapsulated Bacteria

Encapsulation method was used to promote MICP to enable self-healing concrete. Self-healing concrete is concrete treated or with the inclusion of self-healing material during mixing process which self-healing material defined as material that restored nearly or all with original functionality after being damaged [35]. The use of bacteria as self-healing agents by promoting MICP for the autonomous remediation of cracks in concrete has been initiated by Jonkers and Schlangen [44]. The idea of having the encapsulation of bacteria along with its nutrient embedded inside the concrete so that the healing process would start from inside to outside of concrete. There might be a risk that the bacterial cells or spores damaged during the mixing stage and cement hydration since continuous hydration gradually hardened the cement-based matrix. Furthermore, damage to bacteria also caused by the mechanical forces during mixing. Therefore, several investigations were conducted of the approach on combining microorganisms and nutrient and encapsulated it together as a reaction component.

The encapsulation method of bacteria also known as a carrier. Wang et al. [93] ascribed the carrier should have a shell function to protect the bacteria, no hindering effect on calcite precipitation upon cracking and shall have no adverse effect on the concrete matrix. The shell should also require minimal external triggers to activate the reaction components of microorganisms and nutrients in the shell matrix to fulfil self-healing function [24, 42, 43, 89, 93]. Besides, the encapsulation of bacteria by means to protect the bacteria with its nutrient so that during mixing process it can distribute homogeneously in the cement-based matrix without destroying the bacterial cell.

In conventional concrete, a minor crack occurs will allow the ingress of water and deleterious material in the crack propagation. Without proper treatment, the continuously ingress water and chemicals in crack propagation would eventually cause premature matrix degradation and corrosion of embedded steel reinforcement [46]. However, the introduction of the bacteria based healing agent during the mixing process would overcome the problem. The embedment of a healing agent which initiated by crack and ingress water would activate the bacteria. The bacterial activity produces copious mineral precipitation that aid in crack closure and further entry of harmful material would be impeded to protect the reinforcement corrosion. The concept of self-healing using bacteria as self helaing agent is illustrated in Fig. 3.4.

A compatible material to encapsulate the bacteria and its nutrients have two (2) distinctive purposes which are; a) the suitability of compatible material with the cement matrix, and b) the appropriateness of incorporated bacteria to survive for prolonged periods of time [24]. The efficiency of crack healing also influenced by the hydration degree and internal structures of concrete at different ages [58].

By encapsulating, the bacteria would be prolonged as a concrete structure maturing, thus continuously enhance the mechanical properties and durability of concrete structures [83].

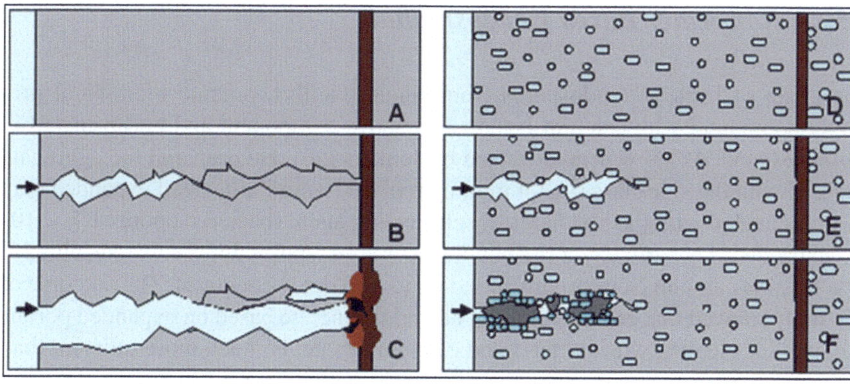


**Fig. 3.4** Schematic drawing of conventional concrete (**a**–**c**) versus bacteria-based self-healing concrete (**d**–**f**). Crack ingress chemicals degrade the material matrix and accelerate corrosion of the reinforcement (**a**–**c**). Incorporated bacteria-based healing agent activated by ingress water seals and prevents further cracking (**d**–**f**) [24]. Reprinted from Ecological Engineering, 36(2), Willem De Muynck, Nele De Belie, Willy Verstraete, Microbial carbonate precipitation in construction materials: A review, 118–136, Copyright 2010, with permission from Elsevier

In conjunction of Jonkers and Schlangen [44] works, many types of research conducted with various types of encapsulation proposed ever since. For instance, Van Tittelboom et al. [90] encapsulated *Bacillus sphaericus* in silica sol-gel matrix. The bacteria can precipitate $CaCO_3$ when encapsulated in the gel rather than without encapsulation and efficiently heal up to 10 mm deep cracks. Crack sealing employing this biological treatment resulted in a decrease in water permeability. However, the decline in water flow also obtained if autoclaved bacteria were used instead of active bacteria. This corroborates that the greater part of the decrease in water permeability is attributed to crack filling by the sol–gel matrix. Thermogravimetric Analysis (TGA) analysis of the crack repair material showed only in the case of active bacteria used proven by the presence of $CaCO_3$ crystals. Precipitation of these crystals in the gel matrix may enhance the durability of this repair material.

The following section will entail on the encapsulation method of bacteria that adopted for MICP in crack remediation.

## 3.5 Encapsulation Methods of Bacteria for Self-Healing in Concrete

The encapsulation method categorised under two (2) categories, namely non-water based encapsulation and water based encapsulation. The review of this approach is explained in the following sub-sections.

### 3.5.1 Non-water Based Encapsulation

The effect of bacterial endospores from genus Bacillus, organic biomineral precursor compound with calcium lactate as reactive component and encapsulated in porous expanded clay was investigated by Jonkers [46]. He prepared the aggregate material with the size class of 2–4 mm, by replaced a similarly sized expanded clay particles loaded with the biochemical self-healing agent (bacterial spores $1.7 \times 10^5$ $g^{-1}$ expanded clay particles, corresponding to $5 \times 10^7$ spores $dm^{-3}$ concrete, plus 5% w/w fraction calcium lactate, corresponding to 15 g $dm^{-3}$ concrete). The outcome of this study showed that crack healing of bacterial concrete based on expanded porous clay particles loaded with bacteria and calcium lactate, is much more efficient than of concrete of the same composition, however, with expanded clay particles. The reason for this can be explained by the strictly chemical processes in the control and additional biological processes in the bacterial concrete. The incorporation of bacterial spores shows the capability of spores in a conversion from calcium lactate to calcium carbonate materials. Improvement in concrete durability due to quantification of self-healing capacity. The encapsulation of bacterial spores able to revive and produce copious minerals after stimulation by a suitable medium. Initial results show that the addition of specific organic mineral precursor compounds plus spore-forming alkaliphilic bacteria as self-healing agents produces up to 100 $\mu$m sized calcite particles which can potentially seal micro to even larger sized cracks. The expanded clay to encapsulate bacteria and biomineral precursor compound as shown in Fig. 3.5.

A similar method as proposed by Jonkers [46] applied in research conducted by Wiktor and Jonkers [100]. Both authors used endospores from *Bacillus alkalinitrilicus* at $1.7 \times 10^5$ bacterial spores $g^{-1}$ which encapsulated in a biomineral precursor compound with calcium lactate. The results revealed that after 100 days of immersion in tap water, bacteria-based mortar specimens had maximum healed

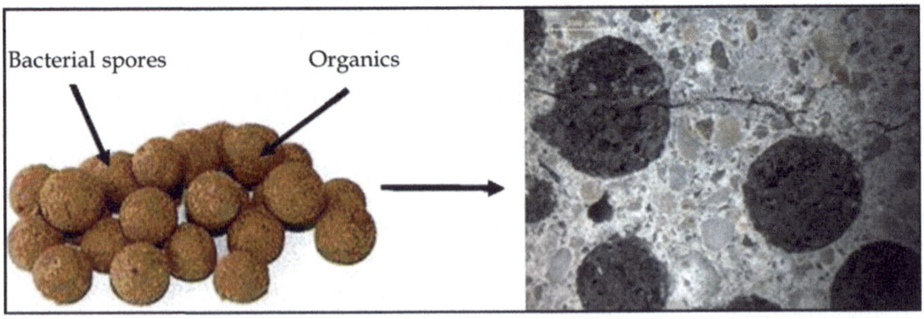


**Fig. 3.5** Expanded clay particles loaded with biomineral precursor compound and bacteria (left), and when embedded in the concrete matrix (right) the loaded expanded clay particles represent internal reservoirs containing the two component healing agents consisting of bacterial spores and a suitable biomineral precursor compound. The size of bacterial spores are 1 μm [46]. Jonkers, H., In-Genium, 1, 84–93, 2021; licensed under a Creative Commons Attribution-NonCommercial-ShareAlike 4.0 International (CC BY-NC-SA 4.0)

of the crack width of more than twice the size (0.46 mm) compared with control specimens (0.18 mm). The enhancement of the self-healing capacity of concrete has promoted the maximum heal able crack width more than doubled, which effectively obtained from a mixture of bacterial spores and calcium lactate.

It is also presumed that due to combined direct and indirect calcium carbonate formation; direct $CaCO_3$ precipitation through metabolic conversion of calcium lactate and indirect formation due to reaction of metabolically produced $CO_2$ molecules with $Ca(OH)_2$ minerals present in the concrete matrix, leading to additional $CaCO_3$ precipitation. With calcium lactate as biomineral precursor compound, Khaliq and Ehsan [48] using the encapsulation method of lightweight aggregate (LWA) and graphite nanoplatelets (GNP) to encapsulate *Bacillus subtilis* spores. To encapsulate *Bacillus subtilis* spores, the LWA and GNP soaked with the bacterial solution before mixing in concrete. Specimens incorporated with graphite nanoplatelets (GNP) as carrier compound displayed uniform distribution and protection of bacteria at samples pre-cracked at an early age of 3 and 7 days, resulting in maximum crack healing efficiency. Consistency in crack healing efficiency in specimens pre-cracked at later days incorporated with lightweight aggregate (LWA). The addition of *Bacillus subtilise* resulted in a slight increase in compressive strength, irrespective of the incorporation technique, with significant improvement through LWA technique. Self-healing concrete with geopolymer coatings as encapsulation method was another approach that has been investigated by de Koster et al. [17]. With *Bacillus* spores, the results showed that the treated geopolymer coating of bacteria containing granules using high-pressure single-fluid nozzle are better than untreated pellets to be used in self-healing concrete. Under low convection, the uncoated healing agent was fully dissolved after two (2) hours as compared to treated one at 65% to 100% of losing coat after three (3) hours. The interaction between cement paste and coating are appeared to be sufficient. All treated cubes gain similar strength indicating no contribution from the healing material and coating.

Cyclic Enriched Ureolytic Powder (CERUP) (Fig. 3.6) has been examined by da Silva et al. [18] as the encapsulation method for housing non-axenic bacterial cultures which comprises of *Sporocarsina pasteurii* and *Bacillus subtilis*. The production of CERUP using cyclic thermal treatment, in which allow the selection of bacterial spores over time. During each cycle measured with 'L' as a scale, one can force the spore forming bacteria to sporulate, allowing them to survive the high temperatures. Using CERUP at 5 L scale revealed good results both concerning urea degradation and calcium carbonate precipitation.

The particular ureolytic activity obtained with CERUP is greater when compared with the one calculated for the axenic cultures. CERUP produced at the 50 L scale can be compared to both ureolytic activity and calcium carbonate precipitation capability with the axenic culture used as a reference. The incorporation of CERUP, either autoclaved or not, increased the autonomous healing of the mortar samples.

Bacterial activity needs water as an essential element. Therefore, to enable bacterial activity in the encapsulation method to be incorporated into cement matrix, water based encapsulation as the carrier is the most suitable for general application. The following sub-section will entail the water-based carrier explored for MICP.

**Fig. 3.6** CERUP after the grinding procedure. Left side: CERUP produced at 5 L scale. Right side: CERUP produced at the 50 L scale [18]. Reprinted from Construction and Building Materials, 93, F. B. da Silva, N. D. Belie, N. Boon, W. Verstraete, Production of non-axenic ureolytic spores for self-healing concrete applications, 1034–1041, Copyright 2015, with permission from Elsevier

### *3.5.2 Water Based Encapsulation*

Wang et al. [97] has initiated the introduction of water-based encapsulation method (carrier) in self-healing concrete. Among the water-based carrier explored for delivering MICP to the target location is sodium alginate gel. Sodium alginate is a polymer which extracted from brown algae [21]. Sodium alginate is one of the hydrogels by natural polymer and requires cross-linker. It acts as bacterial carrier and enzymes due to its biocompatibility in many applications as reported by Dong et al. [26], Dey and Roy [25], and Viveganandan and Jauhri [91]. Sodium alginate has been used for multi purposes from biomedical sciences to tissue engineering due to its diversity and compatibility of multiple purposes [21, 47, 86]. For instance, Park and Lee [69] used alginate hydrogel as cartilage regeneration in tissue engineering. Alginate hydrogel as natural materials widely used in the biomedical field as potential protein delivery dressings to wounds and promote wound healing [6, 20, 62] and live probiotic bacteria provision in the gastrointestinal tract [57, 80]. It has also widely exploited as drug delivery applications [6, 53, 61, 71, 103] due to its capability and potential in the encapsulating material. The advantage of using alginate as cell encapsulation is their insufficient inherent cell adhesion and cellular interaction, a rapid, non-toxic, and versatile method for immobilisation of macromolecules and cells [8]. Alginate is known to form a hydrogel in the presence of divalent cations, such as calcium ($Ca^{2+}$) [51], which act as cross-linkers between the functional groups of alginate chains [104]. The divalent cations form an alginate gel structure bind solely the blocks of the alginate chains, plugging one polymer and formed junctions with

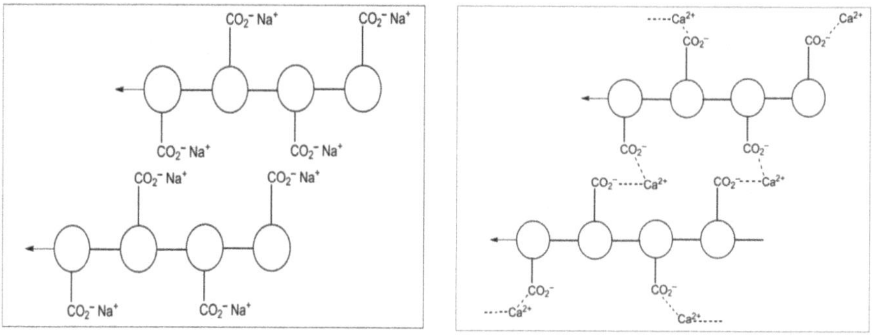


**Fig. 3.7** The polymer chains of sodium alginate:; **a** before contacted with calcium chloride and **b** when contacted with calcium chloride [70]. Reprinted with permission from Pignolet, L. H., Waldman, A. S., Schechinger, L., Govindarajoo, G., & Nowick, J. S. (1998). The Alginate demonstration: polymers, food science, and ion exchange. Journal of Chemical Education, 75 (11), 1430–1431. Copyright 1998 American Chemical Society

the other blocks from adjacent polymer [51]. The polymer chains of sodium alginate before contacted with cross-linker is presented in Fig. 3.7a while Figure 3.7b shows that each calcium ion can attach to two of the polymer strands when contacted with cross-linker. This process is rapid and irreversible binding reaction. The broad application of sodium alginate and its gelling to hardened properties when contacted with calcium ($Ca^{2+}$) become a potential in encapsulating bacteria for dispersion during the concrete mixing process. Directly incorporation of bacteria into mortar mix not only have a high tendency on diminishing of bacteria due to concrete pore reduction over time, but it also exposed to the high alkaline environment presence in concrete.

As proposed by Jonkers and Schlangen [44] and Zheng et al. [107], the encapsulation as a bacterial carrier was needed, prior addition to the concrete mixture. The first research initiating sodium alginate as a bacterial carrier conducted by Wang et al. [97] which the bacteria and its nutrient encapsulated in the sodium alginate with the combination of agar. The results revealed that about 100 μm deposited a layer of $CaCO_3$ bound tightly with the surface of cement pastes after seven (7) days of treatment.

In conjunction of Wang et al. [97] works, the application of sodium alginate as encapsulation method in the cementitious material matrix has been reported in a research study by Wang et al. [95], Palin et al. [67], Wong [101], Annamalai et al. [9], and Harbottle et al. [36] to heal the cracks. Several ingredients and the different cross-linker agent have been used to prepare the alginate hydrogels used to repair the cracks and as shown in Table 3.6.

On the other side, hydrogels are hydrophilic polymer networks which may absorb from 10 to 20% (an arbitrary lower limit) up to thousands of times their dry weight in water. The character of the water in a hydrogel can determine the overall permeation of nutrients into and cellular products out of the gel as cited by Hoffman [39]. Water based encapsulation method also has been explored due to the needs of the presence

**Table 3.6** The different ingredients and cross-linker agents of alginate hydrogel used for bacterial carrier in self-healing concrete

| Name | Composition | | Types of bacteria | References |
|---|---|---|---|---|
| | Ingredients | Crosslinker | | |
| Microcapsules | Sodium alginate | Calcium chloride | *Bacillus pseudofirmus* | Liu et al. [55] |
| | Chitosan | Glucono-Delta-Lactone (GDL) | | |
| Alginate sodium (AS) hydrogel | Aluminium (Al) | Calcium carbonate | *Sporocarsina pasteurii* | Nassirian et al. [65] |
| | Sodium (Na) | | | |
| Polyurethane | Deionised water (5 ml) | Calcium chloride | *Sporocarsina pasteurii* | Annamalai et al. [9] |
| | Sodium alginate (5 g) | | | |
| Biological gel based | Sodium alginate (3.2%) | Calcium chloride | *Sporocarsina pasteurii* | Harbottle et al. [36] |
| | Nutrient broth (3 g/L) | | | |
| | Urea (20 g/L g) | | | |
| | $NaHCO_3$ (2.1 g/L g) | | | |
| | $NH_4CL$ (10 g/L g) | | | |
| Bio based healing agent | Sodium alginate (2%) | Calcium acetate | Undisclosed | Palin et al. [67] |
| | Magnesium acetate | | | |
| Methacrylate modified alginate | Sodium alginate (20 g) | Irgacure®2959 | *Bacillus sphaericus* | Wang et al. [95] |
| | Methacrylic anhydrate | | | |
| | NaOH | | | |

of water enable bacterial activity. Wang et al. [93] and Wang et al. [94] among the earliest researchers that used hydrogel as a water-based carrier. *Bacillus sphaericus* spores encapsulated in hydrogels along with its nutrients. The hydrogel-encapsulated spores had the carbonate-precipitating capacity to form $CaCO_3$ in/on the hydrogel matrix. The crack of a width 0.5 mm can be completely healed in the specimens with embedded hydrogels loaded with bacteria and nutrients.

The selection of sodium alginate hydrogel for a carrier comes along with those highlighted advantages. The standard tests which involving biochemical analysis and performance test will be conducted to examine the suitability and characteristic of the carrier to be used in self-healing mortar. The biochemical test performed prior encapsulation of bacteria. This test was carried out to ensure the ability of bacteria to produce urease that helps in the formation of calcium carbonate. Another test is an indication of calcium carbonate which performed to the bacteria by supplying the nutrient that will aid bacteria to produce calcium carbonate. Meanwhile, the performance test carried out after encapsulation of bacteria. The various testings conducted to evaluate the encapsulated bacteria and performance described in Table 3.7.

There are four (4) tests involved in the performance test namely swelling properties, water retention properties, urease activity of bacteria and also viability analysis of bacteria. For instance, using the water-based encapsulation method need specific performance test such as swelling test and water retention properties. The results of swelling properties, water retention properties, urease activity and also viability analysis of *Geobacillus stearothermophilus* encapsulated with sodium alginate hydrogel from the experimental works performed by the authors are elaborated in Chap. 5.

Swelling capacity as defined by Yang [105] is the required parameter as it indicates the ability of the hydrogel to swell rather that dissolved when water or solvent enters it. However, the consideration of the excessive swelling rate rendered as it will cause the hydrogel to break quickly with the sudden release of the encapsulated material before reaching targeted/specific location [16]. In the other hand, water retention properties measure the ability of the hydrogel to retain its water [98] so that it acts as a water reservoir for bacteria to perform its activity [95].

The suitability of bacteria in the encapsulation material measured by the urease activity of bacteria and viability test. The increased activity in the encapsulation material indicates higher production of urease. According to Wang et al. [95]; Wang et al. [97], urease activity also portrayed the suitability of selected bacteria to perform a bacterial activity, thus, producing urease that can be measured quantitatively. Cell concentration of bacteria plays a significant role in calcite precipitation. Therefore, viability test is conducted to quantify the number of cells before and after the encapsulation process. According to Callone et al. [14], bacteria would diminish under mechanical and shear stresses. The reduction number would differ with the types of bacteria and the process conducted. Hence, this test is an important measure so that the intent number of cell concentration to perform a bacterial activity in self-healing concrete are achievable.

**Table 3.7** The overview of the different testing types on water based encapsulation used for bacterial carrier in self healing concrete

| Group | Testing types | Purpose | References |
|---|---|---|---|
| Biochemical test | Indication of urease production | Ensure bacteria capability in metabolic conversion of biomineralisation | Andalib et al. [7], Annamalai et al. [9] |
| | Indication of calcium carbonate precipitation | Ability of bacteria to produce calcium carbonate under supplied nutrient | Annamalai et al. [9] |
| Performance test | Swelling properties | To measure the water absorption capacity by the hydrogel | Nassirian et al. [65], Wang et al. [95], Palin et al. [67], Wang et al. [97] |
| | Water retention properties | To test the moisture uptake by the hydrogel | Wang et al. [95], Nassirian et al. [65] |
| | Urease activity of bacteria | To measure whether encapsulated bacteria in hydrogel able to germinate and utilise supplied nutrient sources | Palin et al. [67] |
| | Viability test of bacteria | To measure the bacteria cell reduction number for pre-encapsulation and post-encapsulation | Wang et al. [95], Nassirian et al. [65] |

## 3.6 Effects of Incorporation of Encapsulated Material as MICP in Concrete

The addition of encapsulated material as MICP has proven giving a positive result on healing effectiveness and efficiency. Nevertheless, the addition of encapsulation material also impart negative effect towards mechanical properties of concrete and has been reported in the literature. For instance, the reduction strength [4, 13, 37, 87, 89, 94] and increase of water permeability [87, 92] has been reported by incorporation of encapsulation material in mortar or in concrete.

Various factors will influence strength such as an amount of encapsulation material incorporated into the concrete mix, water to cement ratio, aggregates grading, age and curing condition also admixtures types that will influence the hydration degree and microstructure of the specimen [94]. In research conducted by Wang et al. [95], the workability of concrete did not affect significantly by the addition of 0.5 and 1.0% by mass of hydrogel; however, with the addition of 1.0% of the hydrogel, the tensile strength and compressive strength declined at about 23.40% and 30% respectively. The formation of macro voids due to the addition of hydrogel created such adverse effect that influences the strength results ([95, 106];). Nevertheless, Wang et al. [95] and Jonkers et al. [45] has reported the existence of macropores in in concrete caused by hydrogels increased the freeze and thaw resistance. In standard practice, the air-entraining agent is used during the mixing process to induce macropores in the final product that able to resist freeze and thaw. The positive effect of freeze and thaw thwart the adverse effect of concrete strength properties. The comparison of each encapsulation material along with the bacterial type listed as in Table 3.8 to observe the effect on encapsulated bacteria in the cementitious material.

The study conducted by Tziviloglou et al. [89] revealed that by incorporating lightweight aggregate (LWA) as sand replacement leads to a substantial decrease of compressive strength at 37.8% as compared to control specimen. Meanwhile, the incorporation of LWA by Jonkers [46] that replace total aggregate volume also experience the reduction of compressive strength at 50% due to a higher percentage of replacement. The incorporation of Cyclic Enriched Ureolytic Powder (CERUP) by da Silva et al. [18] at 3 and 5% (of the cement weight) decreased the compressive strength of the mortar specimens considerably.

However, incorporation of CERUP at 0.5% and 1% (of the cement weight) did not affect the behaviour of the mortar specimens drastically. The researchers concurred that higher percentage of CERUP was not suitable in benefiting mechanical properties and suggesting the incorporation of CERUP maximum at 1% from total weight of the cement.

Reduction of compressive strength also observed on the integration of *Bacillus sphaericus* in three (3) different encapsulation methods namely metakaolin (MK), zeolite (Z) and air-entrainment (AE) at 66.6%, 16.4% and 65.4%, respectively [29]. Tziviloglou et al. [89] concluded such high decreases due to the formation of multiple air voids which localised at the bottom specimens as a result of foam formation of protein dissolution by bacteria nutrient which embedded along with encapsulation material. However, a different result obtained in a study conducted by Erşan et al. [29] and Khaliq and Ehsan [48], showing that as the mortar age increased, the strength obtained also increased. In an experimental work carried out by Erşan et al. [29], which used encapsulated *Diaphorobacter nitroreducens* in three (3) different encapsulation materials namely diatomaceous earth (DE), expanded clay (EC) and granular activated carbon (GAC). Positive results on compressive strength obtained by increasing value of 2.2%, 15.7% and 21.2%, respectively. On the other hand, Khaliq and Ehsan [48] with specimens contained lightweight aggregate (LWA) showed compressive strength was increased to 12% while graphite nanoplatelets (GNP) at 9.8% as compared to control specimens. A small percentage of inclusion of LWA and GNP which less than 1% contribute to this result. Such improvement shows that the suitability of the bacteria used along with bonding with mortar matrix as a contributing factor that giving significant improvement in compressive strength. Flexural strength was also another measure to determine the suitability and performance of mortar incorporated with various encapsulation method. Table 3.9 summarises the

**Table 3.8** Bacteria used, encapsulation methods, weight, cell concentration, compressive strength and improvement made by incorporating the self healing agents in concrete reported by researchers

| References | Bacteria | Encapsulation method, weight | Cell concentration (cfu/ml) | Compres-sive strength at 28 days (MPa) | % increase with reference to control specimens |
|---|---|---|---|---|---|
| Khaliq and Ehsan [48] | *Bacillus subtilis* | Control | – | 26.3* | – |
| | | LWA, 0.07% | $3 \times 10^8$ | 29.4* | 11.8 |
| | | GNP, 0.07% | $3 \times 10^8$ | 28.9* | 9.9 |
| Tziviloglou et al. [89] | *Bacillus* | Control | – | 45.0* | – |
| | | LWA, 16.75 | $1 \times 10^8$ Spores/L | 28.0* | −37/8 |
| da Silva et al. [18] | Non-axenic bacterial culture | Control | – | 57.0* | – |
| | | CERUP, 0.5% | ND | 56.0* | −1.8 |
| | | CERUP, 1% | ND | 53.0* | −7.0 |
| | | CERUP, 3% | ND | 37.0* | −35.1 |
| | | CERUP, 5% | ND | 28.0* | −50.9 |
| Erşan et al. [29] | *Diaphorobacter nitroreducens* | Control | – | 58.6 | – |
| | | DE, 5% | – | 59.9 | 2.2 |
| | | EC, 5% | – | 67.8 | 15.7 |
| | | GAC, 5% | – | 71.0 | 21.2 |
| | *Bacillus sphaericus* | MK, 5% | – | 19.6 | −66.6 |
| | | Z, 5% | – | 49.0 | −16.4 |
| | | AE, 5% | – | 20.3 | −65.4 |
| Wang et al. [95] | *Bacillus sphaericus* | Control | – | 65.0* | – |

(continued)

**Table 3.8** (continued)

| References | Bacteria | Encapsulation method, weight | Cell concentration (cfu/ml) | Compres-sive strength at 28 days (MPa) | % increase with reference to control specimens |
|---|---|---|---|---|---|
| | | AM-H, 0.5% | $2.3 \times 10^9$ Spores/L | 54.0* | −16.9 |
| | | AM-H, 1% | $2.3 \times 10^9$ Spores/L | 50.0* | −23.1 |

ND: Not described

* Interpolated and derived from graph

flexural strength of concrete incorporated various encapsulation methods reported by researchers at 28 days.

The application of lightweight aggregate (LWA) for flexural strength conducted By Tziviloglou et al. [89]. As explained by Tziviloglou et al. [89], the higher incorporation of LWA impregnated with the healing agent (B) at 60% reduced flexural strength as compared to 55% LWA without the healing agent (CTRL) from the total weight of cement.

This results revealed that the healing agent does not contribute to flexural strength enhancement. Wang et al. [95] conducted flexural strength of mortar with the incorporation of methacrylate modified alginate without spore (AM-H). The result shows that the inclusion of AM-H at 0.5% and 1% from total weight of cement reduced flexural strength at 15.6% and 28.6%, respectively. As obtained by Sixuan [84], the incorporation of graphite nanoplatelets (GNP) at 1% and 1.5%, decreased the flexural strength at 4.7% and 10.6%, with respect to control mortar. To summarise, those results delineated the incorporation of encapsulated material at high percentage would pronounce negative effect on flexural strength.

Furthermore, this drawback also occurred due to macropore formation by encapsulation material as they take up mixing water, and after hardening process the water releases process takes place; remaining an empty macropore; as a result [41, 50, 59, 85]. The existence of pore influences the strength development of concrete. For instance, the total amount of air voids and macropores in the cross-sectional area of the tensile plane would impact the bending strength of the concrete [95]. In the other hand, composition and the microstructure of the specimen also influenced the strength in which further hydration was causing more pore space to be taken by solid phase; thus less porosity and higher strength will result. Therefore, the microstructure is the 'root' factor influencing strength due to an inclusion of encapsulated material that changes the character of a microstructure.

In contrast, the addition of encapsulated material has reported decreasing the rate of water absorption. Gupta et al. [34] revealed that premature cracking or breaking of encapsulated material would occur during the mixing process. This broken encapsulated material reduced the pore size which indicated smaller space taken by broken encapsulated material. However, space which is made by encapsulated material can still be regarded as an orifice or weak point during strength testing, since it cannot withstand the load from compression in contradict with the condition during the process of water absorption in which the encapsulated material will stay intact and block the pores. This is one of the reasons why the lower porosity did not contribute to a higher strength in the specimens with encapsulated material.

Chapter 5 presents further the characteristics of *Geobacillus stearothermophilus* encapsulated with sodium alginate hydrogel and immersed in cement slurry and water.

**Table 3.9** Bacteria used, encapsulation methods, weight, cell concentration, flexural strength and improvement made by incorporating the self healing agents reported by previous researchers

| References | Bacteria | Encapsulation method, weight | Cell concentration (cfu/ml) | Flexural strength at 28 days (MPa) | % increase with reference to control specimens |
|---|---|---|---|---|---|
| Tziviloglou et al. [89] | *Bacillus* | Control | – | 7.0* | – |
| | | CTRL, 55% | – | 6.3* | −10 |
| | | B, 60% | – | 5.5* | −21.4 |
| Wang et al. [95] | *Bacillus* | Control | – | 7.7* | – |
| | | AM-H, o.5% | – | 6.5* | −15.6 |
| | | AM-H, 1% | – | 5.5* | −28.6 |
| Sixuan [84] | Non-axenic bacterial culture | Control | – | 8.5 | – |
| | | GNP, 1% | – | 8.1 | −4.7 |
| | | GNP, 1.5% | – | 7.6 | −10.6 |

* Interpolated and derived from graph

## References


1. Achal, V., Pan, X.: Characterization of urease and carbonic anhydrase producing bacteria and their role in calcite precipitation. Curr. Microbiol. **62**(3), 894–902 (2011). https://doi.org/10.1007/s00284-010-9801-4
2. Achal, V., Mukherjee, A., Reddy, M.: Microbial concrete: way to enhance the durability of building structures. J. Mater. Civ. Eng. **23**(6), 730–734 (2011a). https://doi.org/10.1061/(ASCE)MT.1943-5533.0000159
3. Achal, V., Pan, X., Özyurt, N.: Improved strength and durability of fly ash-amended concrete by microbial calcite precipitation. Ecol. Eng. **37**(4), 554–559 (2011b)
4. Ahmed, M., Mallick, J., Hasan, M.A.: A study of factors affecting the flexural tensile strength of concrete. J. King Saud Univ. Eng. Sci. **28**(2), 147–156 (2014)
5. Almusallam, A., Khan, F., Dulaijan, S., Al-Amoudi, O.: Effectiveness of surface coatings in improving concrete durability. Cem. Concr. Compos. **25**(4), 473–481 (2003)
6. Al-Roujayee, A.S., Hilaj, E., Deepak, A., Jyothi, S.R., Hamid, J.A., Ariffin, I., Saraswat, V., Garg, A.: Alginate-based systems: advancements in drug delivery and wound healing. Int. J. Polym. Mater. Polym. Biomater. **74**(9), 846–874 (2025)
7. Andalib, R., Muhd Zaimi, A.M., Mohd Warid, H., Mohanadoss, P., Ali, K., Jahangir, M., Han-Seung, L.: Optimum concentration of *Bacillus megaterium* for strengthening structural concrete. Constr. Build. Mater. **118**, 180–193 (2016). https://doi.org/10.1016/j.conbuildmat.2016.04.142
8. Andersen, T., Auk-Emblem, P., Dornish, M.: 3D cell culture in alginate hydrogels. Microarrays **4**(2), 133–161 (2015)
9. Annamalai, S.K., Arunachalam, K.D., Sathyanarayanan, K.S.: Production and characterization of bio caulk by *Bacillus pasteurii* and its remediation properties with carbon nano tubes on concrete fractures and fissures. Mater. Res. Bull. **47**(11), 3362–3368 (2012). https://doi.org/10.1016/j.materresbull.2012.07.024
10. Bačun-Družina, V., Butorac, A., Mrvčić, J., Landeka Dragičević, T., Stehlik-Tomas, V.: Bacterial stationary-phase evolution. Food Technol. Biotechnol. **49**(1), 13–23 (2011)
11. Bang, S.S., Galinat, J.K., Ramakrishnan, V.: Calcite precipitation induced by polyurethane-immobilized *Bacillus pasteurii*. Enzym. Microb. Technol. **28**(4–5), 404–409 (2001). https://doi.org/10.1016/s0141-0229(00)00348-3
12. Bang, S.S., Lippert, J.J., Yerra, U., Mulukutla, S., Ramakrishnan, V.: Microbial calcite, a bio-based smart nanomaterial in concrete remediation. Int. J. Smart Nano Mater. **1**(1), 28–39 (2010). https://doi.org/10.1080/19475411003593451
13. Biswas, M., Majumdar, S., Chowdhury, T., Chattopadhyay, B., Mandal, S., Halder, U., Yamasaki, S.: Bioremediase a unique protein from a novel bacterium BKH1, ushering a new hope in concrete technology. Enzym. Microb. Technol. **46**(7), 581–587 (2010). https://doi.org/10.1016/j.enzmictec.2010.03.005
14. Callone, E., Campostrini, R., Carturan, G., Cavazza, A., Guzzon, R.: Immobilization of yeast and bacteria cells in alginate microbeads coated with silica membranes: procedures, physico-chemical features and bioactivity. J. Mater. Chem. **18**(40), 4839–4848 (2008)
15. Chahal, N., Siddique, R., Rajor, A.: Influence of bacteria on the compressive strength, water absorption and rapid chloride permeability of fly ash concrete. Constr. Build. Mater. **28**(1), 351–356 (2012). https://doi.org/10.1016/j.conbuildmat.2011.07.042
16. Cook, M.T., Tzortzis, G., Charalampopoulos, D., Khutoryanskiy, V.V.: Production and evaluation of dry alginate-chitosan microcapsules as an enteric delivery vehicle for probiotic bacteria. Biomacromolecules **12**(7), 2834–2840 (2011)
17. de Koster, S.A.L., Mors, R.M., Nugteren, H.W., Jonkers, H.M., Meesters, G.M.H., van Ommen, J.R.: Geopolymer coating of bacteria-containing granules for use in self-healing concrete. Procedia Eng. **102**, 475–484 (2015). https://doi.org/10.1016/j.proeng.2015.01.193
18. da Silva, F.B., De Belie, N., Boon, N., Verstraete, W.: Production of non-axenic ureolytic spores for self-healing concrete applications. Constr. Build. Mater. **93**, 1034–1041 (2015). https://doi.org/10.1016/j.conbuildmat.2015.05.049

19. Dang, Y., Xie, N., Kessel, A., McVey, E., Pace, A., Shi, X.: Accelerated laboratory evaluation of surface treatments for protecting concrete bridge decks from salt scaling. Constr. Build. Mater. **55**, 128–135 (2014)
20. Davidson, J.R.: Current concepts in wound management and wound healing products. Vet. Clin. North Am. Small Anim. Pract. **45**(3), 537–564 (2015). https://doi.org/10.1016/j.cvsm.2015.01.009
21. Dekamin, M.G., Peyman, S.Z., Karimi, Z., Javanshir, S., Naimi-Jamal, M.R., Barikani, M.: Sodium alginate: an efficient biopolymeric catalyst for green synthesis of 2-amino-4h-pyran derivatives. Int. J. Biol. Macromol. **87**, 172–179 (2016)
22. De Muynck, W., Cox, K., De Belie, N., Verstraete, W.: Bacterial carbonate precipitation as an alternative surface treatment for concrete. Constr. Build. Mater. **22**(5), 875–885 (2008a). https://doi.org/10.1016/j.conbuildmat.2006.12.011
23. De Muynck, W., Debrouwer, D., De Belie, N., Verstraete, W.: Bacterial carbonate precipitation improves the durability of cementitious materials. Cem. Concr. Res. **38**(7), 1005–1014 (2008b)
24. De Muynck, W., De Belie, N., Verstraete, W.: Microbial carbonate precipitation in construction materials: a review. Ecol. Eng. **36**(2), 118–136 (2010). https://doi.org/10.1016/j.ecoleng.2009.02.006
25. Dey, K., Roy, P.: Degradation of chloroform by immobilized cells of *Bacillus* sp. in calcium alginate beads. Biotechnol. Lett. **33**(6), 1101–1105 (2011)
26. Dong, Y., Zhang, Y., Tu, B., Miao, J.: Immobilization of ammonia-oxidizing bacteria by calcium alginate. Ecol. Eng. **73**, 809–814 (2014)
27. Elgendy, M., Nehal, E.E., Hoda, M.S., Gehad, O.D., Mervt, M.A., Kamel, M., Galal, Y., Yehia, A.-G.M.: Bacteria-powered self-healing concrete: breakthroughs, challenges, and future prospects. J. Ind. Microbiol. Biotechnol. **52** (2025). https://doi.org/10.1093/jimb/kuae051
28. Eisa, A.M., Tahwia, A.M., Osman, Y.A., Elemam, W.E.: Characteristics of bacteria based self-healing rubberized concrete for sustainable and durable construction. Sci. Rep. **15**(1), 14758 (2025). https://doi.org/10.1038/s41598-025-97174-1
29. Erşan, Y.C., da Silva, F.B., Boon, N., Verstraete, W., De Belie, N.: Screening of bacteria and concrete compatible protection materials. Constr. Build. Mater. **88**, 196–203 (2015). https://doi.org/10.1016/j.conbuildmat.2015.04.027
30. Ghosh, P., Mandal, S., Chattopadhyay, B.D., Pal, S.: Use of microorganism to improve the strength of cement mortar. Cem. Concr. Res. **35**(10), 1980–1983 (2005). https://doi.org/10.1016/j.cemconres.2005.03.005
31. Ghosh, P., Mandal, S., Pal, S., Bandyopadhyaya, G., Chattopadhyay, B.D.: Development of bioconcrete material using an enrichment culture of novel thermophilic anaerobic bacteria. Indian J. Exp. Biol. **44**(4), 336 (2006)
32. Ghosh, S., Biswas, M., Chattopadhyay, B.D., Mandal, S.: Microbial activity on the microstructure of bacteria modified mortar. Cem. Concr. Compos. **31**(2), 93–98 (2009). https://doi.org/10.1016/j.cemconcomp.2009.01.001
33. Gollapudi, U., Knutson, C., Bang, S., Islam, M.: A new method for controlling leaching through permeable channels. Chemosphere **30**(4), 695–705 (1995)
34. Gupta, S., Pang, S.D., Kua, H.W.: Autonomous healing in concrete by bio-based healing agents—a review. Constr. Build. Mater. **146**, 419–428 (2017). https://doi.org/10.1016/j.conbuildmat.2017.04.111
35. Hager, M.D., Greil, P., Leyens, C., van der Zwaag, S., Schubert, U.S.: Self-healing materials. Adv. Mater. **22**(47), 5424–5430 (2010)
36. Harbottle, M.J., Zhang, J., Gardner, D.R.: Combined physical and biological gel-based healing of cementitious materials. In: ICSHM 2013: Proceedings of the 4th International Conference on Self-Healing Materials. Ghent, Belgium, June 16–20, 2013 (2013)
37. Hasholt, M.T., Jensen, O.M., Kovler, K., Zhutovsky, S.: Can superabsorent polymers mitigate autogenous shrinkage of internally cured concrete without compromising the strength? Constr. Build. Mater. **31**, 226–230 (2012)

38. Helmus, R.A., Liermann, L.J., Brantley, S.L., Tien, M.: Growth advantage in stationary-phase (GASP) phenotype in long-term survival strains of *Geobacter sulfurreducens*. FEMS Microbiol. Ecol. **79**(1), 218–228 (2012)
39. Hoffman, A.S.: Hydrogels for biomedical applications. Adv. Drug Deliv. Rev. **64**, 18–23 (2012)
40. Ibrahim, M., Al-Gahtani, A., Maslehuddin, M., Dakhil, F.: Use of surface treatment materials to improve concrete durability. J. Mater. Civ. Eng. **11**(1), 36–40 (1999)
41. Jensen, O.M., Hansen, P.F.: Water-entrained cement-based materials: II. Experimental observations. Cem. Concr. Res. **32**(6), 973–978 (2002)
42. Jonkers, H.M., Schlangen, E.: Crack repair by concrete immobilized bacteria. In: Proceedings of the First International Conference on Self Healing Materials, pp. 1–7. Springer, Cham (2007a)
43. Jonkers, H.M., Schlangen, E.: Self-healing of cracked concrete: a bacterial approach. In: Proceedings of FRACOS6: Fracture Mechanics of Concrete and Concrete Structures, pp. 1821–1826. Catania, Italy (2007b)
44. Jonkers, H., Schlangen, E.: Development of a bacteria-based self healing concrete. In: Proceedings International FIB Symposium, vol. 1, pp. 425–430 (2008). ISBN: 978-1-4398-2841-0
45. Jonkers, H., Thijssen, A., Muyzer, G., Copuroglu, O., Schlangen, E.: Application of bacteria as self-healing agent for the development of sustainable concrete. Ecol. Eng. **36**(2), 230–235 (2010). https://doi.org/10.1016/j.ecoleng.2008.12.036
46. Jonkers, H.: Bacteria-based self-healing concrete. Heron **56**(1/2) (2011)
47. Khalilah, A., Shuhaimi, M., Rosfarizan, M., Arbakariya, A., Yazid, A.: Optimization of fish gelatin-alginate-genipin as encapsulating matrices for probiotic application using FCCD-RSM. In: IEEE Symposium on Humanities, Science and Engineering Research (SHUSER), pp. 1283–1288 (2012)
48. Khaliq, W., Ehsan, M.B.: Crack healing in concrete using various bio influenced self-healing techniques. Constr. Build. Mater. (Part 1), 102, 349–357 (2016). https://doi.org/10.1016/j.conbuildmat.2015.11.006
49. Krishnapriya, S., Venkatesh Babu, D.L., Prince Arulraj, G.: Isolation and identification of bacteria to improve the strength of concrete. Microbiol. Res. **174**, 48–55 (2015). https://doi.org/10.1016/j.micres.2015.03.009
50. Laustsen, S., Hasholt, M.T., Jensen, O.M.: Void structure of concrete with superabsorbent polymers and its relation to frost resistance of concrete. Mater. Struct. **48**(1–2), 357–368 (2015)
51. Lee, K.Y., Mooney, D.J.: Alginate: properties and biomedical applications. Prog. Polym. Sci. **37**(1), 106–126 (2012). https://doi.org/10.1016/j.progpolymsci.2011.06.003
52. Leitner, D.R., Zingl, F. G., Morano, A.A., Zhang, H., Waldor, M.K.: The Mla pathway promotes Vibrio cholerae re-expansion from stationary phase. mBio **16**(2), e03433-03424 (2025)
53. Li, J., Jiang, C., Lang, X., Kong, M., Cheng, X., Liu, Y., Feng, C., Chen, X.: Multilayer sodium alginate beads with porous core containing chitosan based nanoparticles for oral delivery of anticancer drug. Int. J. Biol. Macromol. **85**, 1–8 (2016). https://doi.org/10.1016/j.ijbiomac.2015.12.064
54. Li, T., Yang, H., Yan, X., He, M., Gu, H., Yu, L.: Surface crack treatment of concrete via nano-modified microbial carbonate precipitation. J. Infrastruct. Preserv. Resil. **5**(1), 2 (2024). https://doi.org/10.1186/s43065-024-00095-y
55. Liu, L., Li, Y., Song, J., Zhou, J., Yi, W., Ge, Y., Gao, K.: Cracks repairing and resistance to water penetration properties of microbial self-healing cement. Eng. Life Sci. **25**(3), e70010 (2025a). https://doi.org/10.1002/elsc.70010
56. Liu, W., Song, J., Zhao, H., Cui, Z., Zhang, S., Qiao, Y., Cao, C., Li, W.: Mechanically stable and superhydrophobic nano-$SiO_2$@silane/silicate coating for enhanced impermeability of mortar. Constr. Build. Mater. **470**, 140541 (2025b). https://doi.org/10.1016/j.conbuildmat.2025.140541

57. Liu, Y., Sun, Y., Sun, L., Wang, Y.: In vitro and in vivo study of sodium polyacrylate grafted alginate as microcapsule matrix for live probiotic delivery. J. Funct. Foods **24**, 429–437 (2016)
58. Luo, M., Qian, C.-X., Li, R.-Y.: Factors affecting crack repairing capacity of bacteria-based self-healing concrete. Constr. Build. Mater. **87**, 1–7 (2015). https://doi.org/10.1016/j.conbuildmat.2015.03.117
59. Lura, P., Durand, F., Loukili, A., Kovler, K., Jensen, O.M.: Compressive strength of cement pastes and mortars with superabsorbent polymers. In: Proceedings of the International RILEM Conference on Volume Changes of Hardening Concrete: Testing and Mitigation, pp. 117–125. Rilem Publications SARL, Lyngby (2006)
60. Maravelaki-Kalaitzaki, P.: Hydraulic lime mortars with siloxane for waterproofing historic masonry. Cem. Concr. Res. **37**(2), 283–290 (2007). https://doi.org/10.1016/j.cemconres.2006.11.007
61. Meena, P., Singh, P., Warkar, S.G.: Degradable pH-sensitive calcium-crosslinked tragacanth gum/β-cyclodextrin/sodium alginate hydrogel microspheres prepared via ionotropic gelation technique for hydrophobic drug delivery. J. Polym. Environ. **33**(2), 928–944 (2025). https://doi.org/10.1007/s10924-024-03449-5
62. Momoh, F.U., Boateng, J.S., Richardson, S.C., Chowdhry, B.Z., Mitchell, J.C.: Development and functional characterization of alginate dressing as potential protein delivery system for wound healing. Int. J. Biol. Macromol. **81**, 137–150 (2015). https://doi.org/10.1016/j.ijbiomac.2015.07.037
63. Moradllo, M.K., Sudbrink, B., Ley, M.T.: Determining the effective service life of silane treatments in concrete bridge decks. Constr. Build. Mater. **116**, 121–127 (2016). https://doi.org/10.1016/j.conbuildmat.2016.04.132
64. Munemo, R., Kruger, J., van Zijl, G.P.A.G.: Surface treatment of 3DPC interlayers with silicate-based solution for enhanced interfacial bonding. Cem. Concr. Res. **196**, 107948 (2025). https://doi.org/10.1016/j.cemconres.2025.107948
65. Nassirian, H., Tasnimi, A.A., Bakhshi, B., Derakhshan-sefidi, M.: Application of alginate sodium-carbonate calcium hydrogel capsules as a carrier of healing bacteria in concrete. Constr. Build. Mater. **488**, 142084 (2025). https://doi.org/10.1016/j.conbuildmat.2025.142084
66. Nosouhian, F., Mostofinejad, D., Hasheminejad, H.: Influence of biodeposition treatment on concrete durability in a sulphate environment. Biosyst. Eng. **133**, 141–152 (2015). https://doi.org/10.1016/j.biosystemseng.2015.03.008
67. Palin, D., Wiktor, V., Jonkers, H.M.: Bacteria-based agent for self-healing marine concrete. In: Proceedings of 5th International Conference on Self-Healing Materials, pp. 2–6. Durham, USA (2015)
68. Park, S.-J., Park, Y.-M., Chun, W.-Y., Kim, W.-J., Ghim, S.-Y.: Calcite-forming bacteria for compressive strength improvement in mortar. J. Microbiol. Biotechnol. **20**(4), 782–788 (2010)
69. Park, H., Lee, K.Y.: Alginate/hyaluronate hydrogels for cartilage regeneration. J. Control. Release **152**, e233–e234 (2011)
70. Pignolet, L.H., Waldman, A.S., Schechinger, L., Govindarajoo, G., Nowick, J.S.: The alginate demonstration: polymers, food science, and ion exchange. J. Chem. Educ. **75**(11), 1430–1431 (1998)
71. Pourazariyan, A., Shahgholi, M., Karimipour, A.: The effect of initial temperature and ZnO nanoparticle volume fractions on the stability of sodium alginate hydrogel nanocomposite using molecular dynamics simulation. Int. Commun. Heat Mass Transfer **162**, 108641 (2025). https://doi.org/10.1016/j.icheatmasstransfer.2025.108641
72. Qiu, J., Tng, D.Q.S., Yang, E.-H.: Surface treatment of recycled concrete aggregates through microbial carbonate precipitation. Constr. Build. Mater. **57**, 144–150 (2014). https://doi.org/10.1016/j.conbuildmat.2014.01.085
73. Qian, C., Ren, L., Xue, B., Cao, T.: Bio-mineralization on cement-based materials consuming $CO_2$ from atmosphere. Constr. Build. Mater. **106**, 126–132 (2016)
74. Ramachandran, S.K., Ramakrishnan, V., Bang, S.S.: Remediation of concrete using microorganisms. ACI Mater. J. **98**(1) (2001)

75. Ravindranath, N.K., Likhit, M.: Self healing material bacterial concrete. In: National Conference on 'Recent Innovations in Engineering and Technology', pp. 656–659. Bidar, Karnataka, India (2014)
76. Reddy, S.S.P., Rao, M.S., Aparna, P., Sasikala, C.: Performance of ordinary grade bacterial (*Bacillus subtilis*) concrete. Int. J. Earth Sci. Eng. **3**(1), 116–124 (2010)
77. Sahoo, K.K., Arakha, M., Sarkar, P., Jha, S.: Enhancement of properties of recycled coarse aggregate concrete using bacteria. Int. J. Smart Nano Mater. 1–17 (2016a)
78. Sahoo, K.K., Sathyan, A.K., Kumari, C., Sarkar, P., Davis, R.: Investigation of cement mortar incorporating *bacillus sphaericus*. Int. J. Smart Nano Mater. 1–15 (2016b)
79. Sangadji, S., Wiktor, V., Jonkers, H., Schlangen, H.: Injecting a liquid bacteria-based repair system to make porous network concrete healed. In: IC- SHM 2013: Proceedings of the 4th International Conference on Self-Healing Materials. Ghent, Belgium, June 16–20, 2013 (2013)
80. Sia, C.S., Bong, W.A., Goh, B.-H., Chong, L.H., Tey, B.T., Low, L.E.: Dual-responsive magnetoliposome-stabilized Pickering emulsion alginate hydrogel beads for oral drug delivery. Int. J. Biol. Macromol. 146291 (2025)
81. Siddique, R., Nanda, V., Kadri, E.-H., Khan, M.I., Singh, M., Rajor, A.: Influence of bacteria on compressive strength and permeation properties of concrete made with cement baghouse filter dust. Constr. Build. Mater. **106**, 461–469 (2016a). https://doi.org/10.1016/j.conbuildmat.2015.12.112
82. Siddique, R., Singh, K., Kunal, Singh, M., Corinaldesi, V., Rajor, A.: Properties of bacterial rice husk ash concrete. Constr. Build. Mater. **121**, 112–119 (2016b)
83. Sierra-Beltran, M.G., Jonkers, H., Schlangen, E.: Characterization of sustainable bio-based mortar for concrete repair. Constr. Build. Mater. **67**, 344–352 (2014)
84. Sixuan, H.: Multifunctional Graphite Nanoplatelets (GNP) Reinforced Cementitious Composites. Master's thesis, National University Singapore (NUS), Singapore (2012). http://scholarbank.nus.edu.sg/handle/10635/34752
85. Snoeck, D., Schaubroeck, D., Dubruel, P., De Belie, N.: Effect of high amounts of superabsorbent polymers and additional water on the workability, microstructure and strength of mortars with a water-to-cement ratio of 0.50. Constr. Build. Mater. **72**, 148–157 (2014)
86. Syahirah, S.N., Anida, Y., Nazirah, M., Ahmad, S.A., Dahalan, F., Khalilah, A.: Optimization of aloe vera and alginate as encapsulating matrices for *Lactobacillus acidophilus* using FCCD-RSM approach. In: IEEE 3rd International Conference on System Engineering and Technology (ICSET), pp. 43–47 (2013)
87. Talaiekhozani, A., Keyvanfar, A., Shafaghat, A., Andalib, R., Abd Majid, M.Z., Fulazzaky, M.A., Mohamad Zin, R., Lee, C.T., Hussin, M.W., Hamzah, N., Marwar, N.F., Haidar, H.I.: A review of self-healing concrete research development. J. Environ. Treat. Tech. **2**(1), 1–11 (2014)
88. Thomas, M., Karasiewicz, J., Nowicka-Krawczyk, P., Olszyński, R.M., Balcerzak, Ł., Frydrych, M., Sztorch, B., Ślosarczyk, A.: Using an innovative bifunctional siloxane to protect cement composite surfaces from biological corrosion. Int. J. Mol. Sci. **26**(11), 5052 (2025)
89. Tziviloglou, E., Wiktor, V., Jonkers, H., Schlangen, E.: Bacteria-based self-healing concrete to increase liquid tightness of cracks. Constr. Build. Mater. **122**, 118–125 (2016). https://doi.org/10.1016/j.conbuildmat.2016.06.080
90. Van Tittelboom, K., De Belie, N., De Muynck, W., Verstraete, W.: Use of bacteria to repair cracks in concrete. Cem. Concr. Res. **40**(1), 157–166 (2010). https://doi.org/10.1016/j.cemconres.2009.08.025
91. Viveganandan, G., Jauhri, K.: Growth and survival of phosphate-solubilizing bacteria in calcium alginate. Microbiol. Res. **155**(3), 205–207 (2000)
92. Wang, J., De Belie, N., Verstraete, W.: Diatomaceous earth as a protective vehicle for bacteria applied for self-healing concrete. J. Ind. Microbiol. Biotechnol. **39**(4), 567–577 (2012). https://doi.org/10.1007/s10295-011-1037-1
93. Wang, J., Snoeck, D., Van Vlierberghe, S., Verstraete, W., De Belie, N.: Application of hydrogel encapsulated carbonate precipitating bacteria for approaching a realistic self-healing in concrete. Constr. Build. Mater. **68**, 110–119 (2014a)

94. Wang, J., Soens, H., Verstraete, W., De Belie, N.: Self-healing concrete by use of microencapsulated bacterial spores. Cem. Concr. Res. **56**, 139–152 (2014b). https://doi.org/10.1016/j.cemconres.2013.11.009
95. Wang, J., Mignon, A., Snoeck, D., Wiktor, V., Van Vliergerghe, S., Boon, N., De Belie, N.: Application of modified-alginate encapsulated carbonate producing bacteria in concrete: a promising strategy for crack self-healing. Front. Microbiol. **6**(1088), 1–14 (2015). https://doi.org/10.3389/fmicb.2015.01088
96. Wang, J., Ersan, Y.C., Boon, N., De Belie, N.: Application of microorganisms in concrete: a promising sustainable strategy to improve concrete durability. Appl. Microbiol. Biotechnol. **100**(7), 2993–3007 (2016)
97. Wang, R., Qian, C., Wang, J., Cheng, L.: Different treated methods of microbiologically deposited $CaCO_3$ layer on hardened cement paste surface. J. Chin. Ceram. Soc. **10**, 008 (2008)
98. Wang, W.B., Wang, A.Q.: Preparation, swelling and water-retention properties of crosslinked superabsorbent hydrogels based on guar gum. Adv. Mater. Res. **96**, 177–182 (2010)
99. Wiktor, V., Jonkers, H.: Field performance of bacteria-based repair system: pilot study in a parking garage. Case Stud. Constr. Mater. **2**, 11–17 (2015)
100. Wiktor, V., Jonkers, H.M.: Quantification of crack-healing in novel bacteria-based self-healing concrete. Cem. Concr. Compos. **33**(7), 763–770 (2011). https://doi.org/10.1016/j.cemconcomp.2011.03.012
101. Wong, L.S.: Microbial cementation of ureolytic bacteria from the genus *Bacillus*: a review of the bacterial application on cement-based materials for cleaner production. J. Clean. Prod. **93**, 5–17 (2015). https://doi.org/10.1016/j.jclepro.2015.01.019
102. Xue, X., Yang, J., Zhang, W., Jiang, L., Qu, J., Xu, L., Zhang, H., Song, J., Zhang, R., Li, Y., Qin, J.: The study of an energy efficient cool white roof coating based on styrene acrylate copolymer and cement for waterproofing purpose—part II: mechanical and water impermeability properties. Constr. Build. Mater. **96**, 666–672 (2015). https://doi.org/10.1016/j.conbuildmat.2015.08.033
103. Yan, J., Miao, Y., Tan, H., Zhou, T., Ling, Z., Chen, Y., Xing, X., Hu, X.: Injectable alginate/hydroxyapatite gel scaffold combined with gelatin microspheres for drug delivery and bone tissue engineering. Mater. Sci. Eng. C **63**, 274–284 (2016)
104. Yang, J.-S., Xie, Y.-J., He, W.: Research progress on chemical modification of alginate: a review. Carbohydr. Polym. **84**(1), 33–39 (2011). https://doi.org/10.1016/j.carbpol.2010.11.048
105. Yang, T.: Mechanical and Swelling Properties of Hydrogels. Doctoral dissertation, KTH Royal Institute of Technology (2012)
106. Zhang, J., Liu, Y., Feng, T., Zhou, M., Zhao, L., Zhou, A., Li, Z.: Immobilizing bacteria in expanded perlite for the crack self-healing in concrete. Constr. Build. Mater. **148**, 610–617 (2017). https://doi.org/10.1016/j.conbuildmat.2017.05.021
107. Zheng, M., Zhang, Y., Yan, S., Wu, Z., Li, X., Wu, D., Xiong, L.: Effect of carrier-encapsulated microbial calcium carbonate on the performance of cement mortar. Constr. Build. Mater. **483**, 141579 (2025). https://doi.org/10.1016/j.conbuildmat.2025.141579
108. Zambrano, M.M., Kolter, R.: Gasping for life in stationary phase. Cell **86**(2), 181–184 (1996)

# Chapter 4
# Evaluation of Self-healing Performance in MICP-Based Concrete

## 4.1 Introduction

Self-healing concrete has emerged as a significant advancement in sustainable construction, particularly through the use of Microbially Induced Calcite Precipitation (MICP) to seal cracks and restore structural integrity. To assess the effectiveness of such healing mechanisms, various evaluation methods have been developed and reported in the literature. These methods are essential for verifying the performance, durability, and practical applicability of self-healing technologies in real-world conditions.

This chapter presents a review of the evaluation techniques used to assess self-healing efficiency in concrete, with a specific focus on those applied to MICP-based systems. The methods are broadly categorised into three main types: **visual inspection**, **morphological analysis**, and **confirmation of calcite precipitation** as the primary crack-filling material. Each category reflects a different aspect of the healing process, from surface-level crack closure to microstructural changes and chemical validation of healing products. The findings from existing studies are reviewed and discussed in the following sections.

## 4.2 Visual Inspection

Multiple ways to visualise healing efficiency have been reported in the literature. The visual evaluation was performed on the repaired cracks using a device to capture the concrete/mortar surface under different names such digital volumetric correlation (DVC) to visually and quantitatively evaluate the crack bonding rate of self-healed mortars [13], Raman Spectroscopy [14], light microscope [35], optical microscope [14], crack-measuring microscope [17], stereomicroscope [37] and photomicrograph [18]. Figures 4.1, 4.2 and 4.3 show the visualisation on the formation of calcium

 
R. M. A. M. Azam and D. H. M. Saman, *Self-Healing Concrete*, Engineering Materials,
https://doi.org/10.1007/978-981-95-7553-4_4

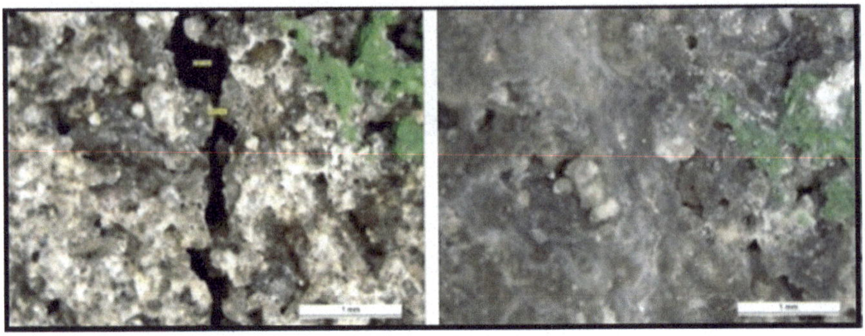

**Fig. 4.1** Maximum crack width healed in the specimen captured using light microscope (initial: 507 μm; final: 0 μm) [35]. Reprinted from Construction and Building Materials, 68, J. Y. Wang, D. Snoeck, S. Van Vlierberghe, W. Verstraete, N. De Belie, Application of hydrogel encapsulated carbonate precipitating bacteria for approaching a realistic self-he aling in concrete, 110–119, Copyright 2014, with permission from Elsevier

carbonate in cracks using light microscope [35], crack-measuring microscope [17] and stereomicroscope [37] respectively.

Likewise, Qian et al. [23] using a digital camera to characterise the surface efflorescence of mortar as shown in Fig. 4.4. The image taken by the digital camera transferred to Photoshop (image software) and the resolution was changed to 8-bit images. The 8-bit images represented on the white area indicating the formation of calcium carbonate as deposited layer on the surface.

Krishnapriya et al. [18] using direct photomicrograph technique that capture crack remediation as shown in Figs. 4.5 and 4.6 respectively. The white precipitations were observed in the cracks as indication of healing due to presence of bacteria.

Figure 4.5 shows the healing proses by *Bacillus megaterium* after 81 days while Fig. 4.6 shows the photomicrograph healing by cyclic enriched ureolytic powder (CERUP) incorporated in mortar specimens.

The benefit of using this method is an easy determination of surface fine structure, the images produced are in natural colour with magnifications [20]. Due to the lightness and easy to set up, the capturing process can be done almost everywhere. It also clearly shows any imperfections that might not be visible to eyes.

## 4.3 Morphological Examination

Scanning Electron Microscopy (SEM) was conducted to obtain sample surface topography and composition. SEM contrast with a visual inspection due to its ability on capturing an image with focus depth and resolving capability (ACEPT 1995). SEM can detect and display around 400× magnification as much detail as using light microscope [26]. SEM does not only can capture the morphology of deposited calcium carbonate but also the bacteria incorporated into the cement-based material as well.

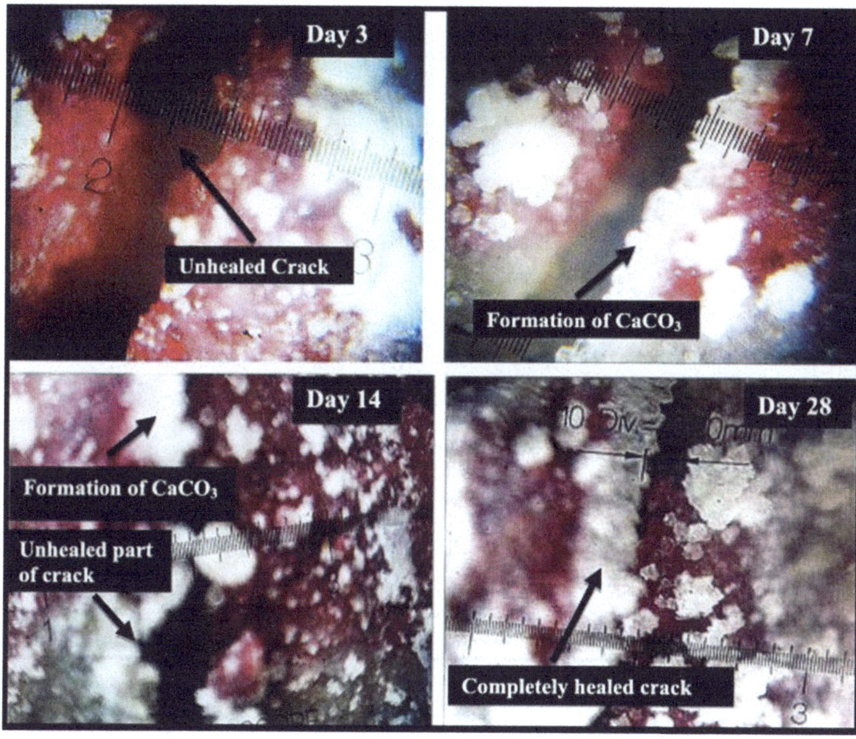


**Fig. 4.2** Formation of calcium carbonate by closer view of cracks captured by crack-measuring microscope (Khaliq et al. 2016). Reprinted from Construction and Building Materials, 102, W. Khaliq, M. B. Ehsan, Crack healing in concrete using various bio influenced self-healing techniques, 349–357, Copyright 2015, with permission from Elsevier

Calcite formation is highly dependent on types of bacteria included as well as the kind of feeding solution supplied to the bacteria. For instance, Wang et al. [32] encapsulated *Bacillus sphaericus* in silica gel with feeding solution of yeast extract and urea. Based on Fig. 4.7, it is noticeable that the $CaCO_3$ particles from silica gel immobilised bacteria had a rhombohedral shape, and the size was about 100 μm. Jonkers and Schlangen [15] captured the calcite formation with needle-like crystal on the specimen's surface after incubated in medium containing yeast extract and peptone as shown in Fig. 4.8. Another calcite shape was observed through SEM image of deposited layer on the specimen surfaces from immobilised bacteria as shown in Fig. 4.9. As seen in Fig. 4.9a), the distribution of the mineralised calcite particles on the mortar specimens. In the other hand, with 5000x magnification from Fig. 4.9a, clear images on the shape of calcite mineralised particles were obtained as seen in Fig. 4.9b.

Figure 4.9b-A displays the mineralised particle that grows independently outside the agar layer. In the other hand, Fig. 4.9b-B characterise the calcite that grows inside the agar layer that was mostly spherical like peanut-shaped [23].

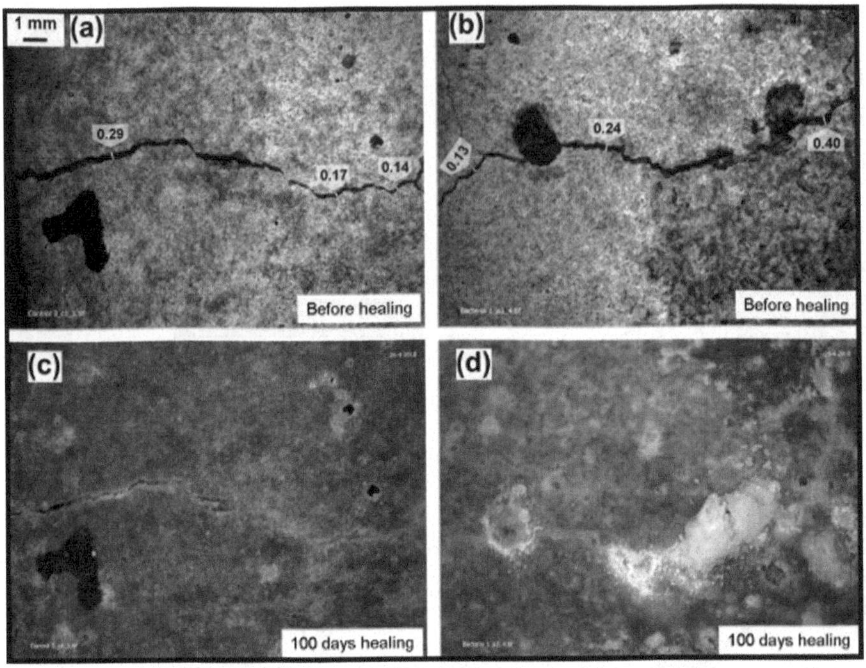


**Fig. 4.3** Stereomicroscopic images of crack-healing process in control mortar specimen before (**a**) and after 100 days healing (**c**); in bio-chemical agent-based specimen before (**b**) and after 100 days of healing (**d**) [37]. Reprinted from Cement and Concrete Composites, 33 (7), Virginie Wiktor, Henk M. Jonkers, Quantification of crack-healing in novel bacteria-based self-healing concrete, 763–770, Copyright 2011, with permission from Elsevier

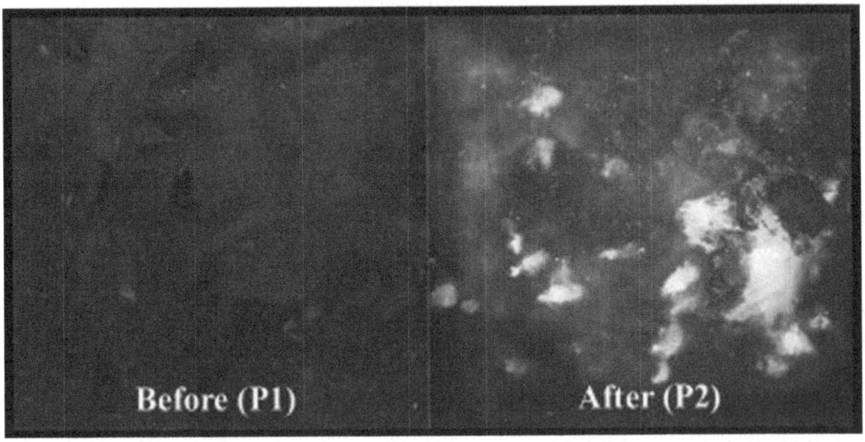


**Fig. 4.4** Mortar wall before and after efflorescence at 8-bit resolution image [23]. Reprinted from Construction and Building Materials, 106, Chunxiang Qian, Lifu Ren, Bin Xue, Tianji Cao, Biomineralisation on cement-based materials consuming $CO_2$ from atmosphere, 126–132, Copyright 2015, with permission from Elsevier

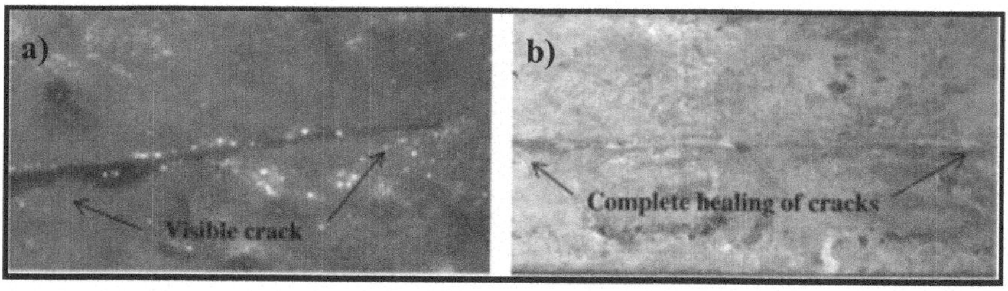


**Fig. 4.5** Photographs of crack healing **a** concrete specimens incorporated with *Bacillus megaterium* before healing and **b** concrete specimens incorporated with *Bacillus megaterium* after 81 days of healing [18]. Reprinted from Microbiological Research, 174, S. Krishnapriya, D. L. Venkatesh Babu, Prince Arulraj G., Isolation and identification of bacteria to improve the strength of concrete, 48–55, Copyright 2015, with permission from Elsevier

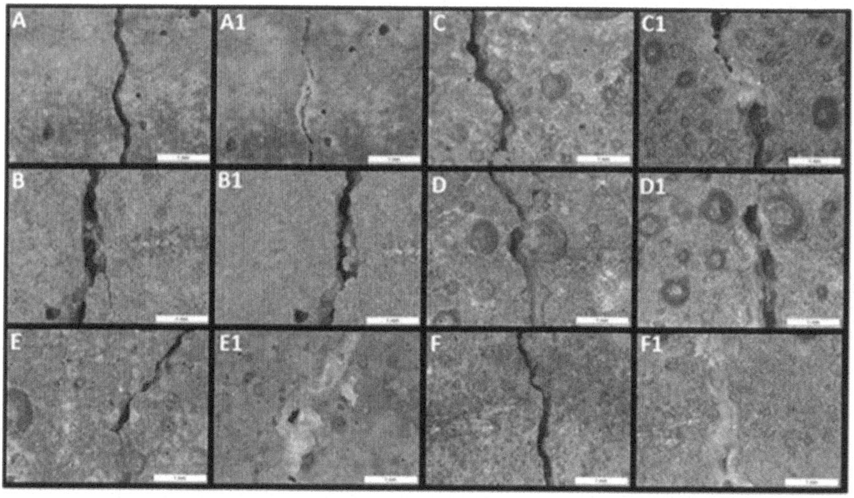


**Fig. 4.6** Photomicrographs of cracked mortar specimens on day 0 and on day 28; **a** and **b** reference samples on day 0; **a1** and **b1** reference samples on day 28; **c** and **d** mortar samples containing 1% of autoclaved CERUP powder on day 0; **c1** and **d1** mortar samples on day 28; **e** and **f** mortar samples containing 1% of CERUP on day 0 and **e1** and **f1** mortar samples on day 28 [9]. Reprinted from Construction and Building Materials, 93, F. B. da Silva, N. D. Belie, N. Boon, W. Verstraete, Production of non-axenic ureolytic spores for self-healing concrete applications, 1034–1041, Copyright 2015, with permission from Elsevier

Different types of calcite morphology observed through SEM examination [39]. This corresponds to the types of bacteria [5], composition of growth media [3], temperature [27], bacteria cell concentration [21] and feeding solutions (nutrient) used [38]. The common types of calcite shape are rhombohedral [36], lamellar [28], spherical (Ma and Feng 2011), [36], orthorhombic [17] and needle-like shapes [24]. However, the most stable form of calcite among all is in rhombohedral shape [8]. SEM

**Fig. 4.7** Scanning electron micrograph (SEM) of rhombohedral shape of $CaCO_3$ precipitation in the silica gel [32]. Reprinted from Construction and Building Materials, 26(1), Jianyun Wang, Kim Van Tittelboom, Nele De Belie, Willy Verstraete, Use of silica gel or polyurethane immobilized bacteria for self-healing concrete, 532–540, Copyright 2011, with permission from Elsevier

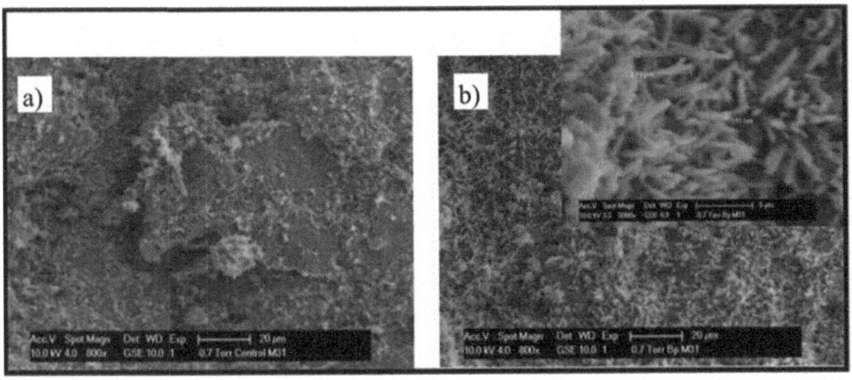


**Fig. 4.8** Concrete samples incubated in yeast extract and peptone containing medium; **a** control (concrete with no bacteria added) and **b** concrete containing $10^9$ cfu/ml of *Bacillus pseudofirmus* endospores. The inset in Fig. 4.8b (5000 magnification) shows a close up of the massive calcite needle-like crystals formed on the concrete surface [15]. Henk M. Jonkers, Self Healing Concrete: A Biological Approach, Self Healing Materials: An Alternative Approach to 20 Centuries of Materials Science, 1, 195–204, 2007, Springer Nature

is also able to analyse specimens microstructure to differentiate the characteristic of samples with and without the presence of bacteria as shown in Fig. 4.10.

As shown in Fig. 4.10a, an image of a porous with large irregularity surface of untreated RCA mortar while Fig. 4.10b, visualise the presence of bacteria change the treated Recycled Concrete Aggregate (RCA) surface through Microbiological Calcium Carbonate Precipitation (MCP) with the composition of spherical shapes crystal. The effect of bacteria presence under different concentration on the concrete porosity also visualised by SEM. As described by Andalib et al. [6], bacterial

**Fig. 4.9** SEM image of deposited layer by immobilised bacteria [23]. Reprinted from Construction and Building Materials, 106, Chunxiang Qian, Lifu Ren, Bin Xue, Tianji Cao, Bio-mineralisation on cement-based materials consuming $CO_2$ from atmosphere, 126–132, Copyright 2015, with permission from Elsevier

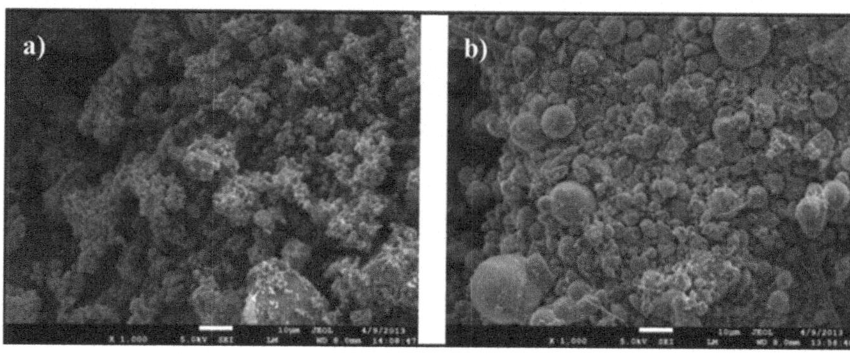


**Fig. 4.10** Scanning electron microscopy (SEM) photos of **a** untreated RCA and **b** MCP-treated RCA [25]. Reprinted from Construction and Building Materials, 57, Jishen Qiu, David Qin Sheng Tng, En-Hua Yang, Surface treatment of recycled concrete aggregates through microbial carbonate precipitation, 144–150, Copyright 2014, with permission from Elsevier

concrete with $10^5$ cfu/ml was filled by a specific mineral (calcite) had less porosity in comparison to other bacterial concentrations as shown in Fig. 4.11.

Apart of that, SEM also manages to capture the imprints of bacteria on the material surface. The details of the bacteria types through imprinted bacteria would determine the suitable bacteria has been embedded/used in the study as contamination would occur without proper handling that would surrogate the evaluation on healing efficiency. The bacterial imprints of *Sporocarsina pasteurii* (*S. Pasteurii*) is as shown in Fig. 4.12.

Further illustrations of morphological examinations using a stereomicroscope and SEM are presented and discussed in the following sections, based on the actual works conducted by the authors.

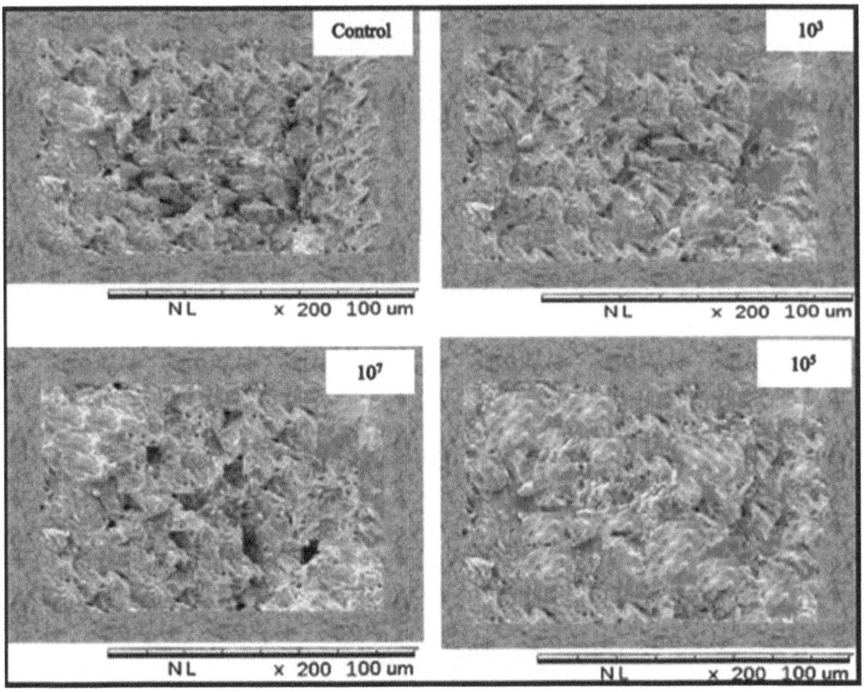


**Fig. 4.11** The comparison of structural concrete porosity in presence of bacteria for $10^3$, $10^5$ and $10^7$ (cfu/ml) concentrations with control specimens under SEM [6]. Reprinted from Construction and Building Materials, 118, Ramin Andalib, Muhd Zaimi Abd Majid, Mohd Warid Hussin, Mohanadoss Ponraj, Ali Keyvanfar, Jahangir Mirza, Han-Seung Lee, Optimum concentration of Bacillus megaterium for strengthening structural concrete, 180–193, Copyright 2016, with permission from Elsevier

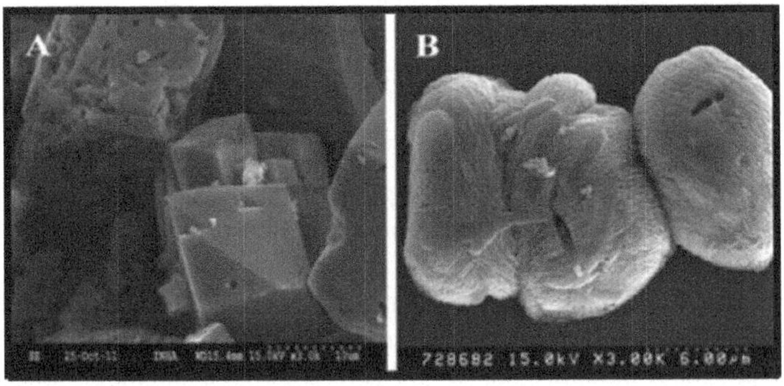


**Fig. 4.12** *Sporocarsina pasteurii* imprints on SEM images of the induced calcium carbonate (A and B) [12]. Gorospe, C. M., Han, S. H., Kim, S. G. et al., Effects of different calcium salts on calcium carbonate crystal formation by Sporosarcina pasteurii KCTC 3558, 18, 903–908, 2013, Springer Nature

## 4.4 Surface Treatment for New Artificial Cracks in Concrete and Mortar

This section Presents an analysis of the healing of the artificial new cracks Created on the Concrete and Mortar Specimens. It Presents the Cracks treated with *S. Pasteurii* and *B.Subtilis* Bacteria Broth Solution. The Broth Solution Was Prepared by Incorporating the bacteria and Differing them According to their growth phases in Order to Obtain the Best Formulation and the most Suitable volume of the Broth to Treat New Cracks. The Analysis Was Assessed Based on the Variables Concerning the duration Taken for the cracks to Be Totally Filled up after the Cracks Were Treated with the Bacteria Broth Solution. The Following Paragraph and Section Explain the method Applying the Bacteria Broth to Crack (surface Treatment) and the Method of Quantifying healing efficiency (HE)

For the treatment of new cracks, several series of concrete and mortar specimens were prepared and artificial cracks were created. Bacteria were selected based on its interval growth for optimum healing efficiency. The bacteria broth solutions were added with different urea concentrations. Then, the injection of the bacteria was performed using a micropipette/dropper with varying amounts of bacteria broth solutions. Different variables were utilised to determine the best bacterium, its composition and the technique of treatment that can provide maximum healing within the shortest time. In the first batch, bacteria taken from three (3) different highest population rates were applied to the artificial cracks of three (3) series of concrete specimens, while no bacteria was not applied to the one (1) control cracked specimen. The number of living cells in the *S. pasteurii*) population over a period of time was examined after 24, 33 and 48 h (Fig. 4.13). Bacterial cell concentrations consisted of 207 $\times$ $10^6$ cfu/ml in 24 h (Specimen 1), 835 $\times$ $10^6$ cfu/ml in 33 h (Specimen 2) and 2.88 $\times$ $10^9$ cfu/ml in 48 h (Specimen 3) were dropped by using a 1 ml pipette/dropper into the crack created on each specimen. Specimens 1 and 2 were taken at intervals of 24 and 33 h after the incubation. Meanwhile, Specimen 3 was taken at the interval of 48 h, which was the highest optical density (OD) measurement implying the highest concentration of bacteria (refer to Table 4.1). The OD was measured using a spectrophotometer (IMPLEN, Germany) (Fig. 4.13).

In the second batch, each amount of 20 mL of urea broth with different urea concentrations (5, 10, 20, 30, 40 and 50 g/L) was inoculated with bacteria culture (OD 0.1600 nm). The cultures were incubated at the desired temperature and placed in an incubator shaker. At 24 h (h) intervals, 1 ml of each culture was sampled, and its turbidity (OD readings) was determined, using Nanophotometer, as shown in Fig. 4.14. Next, the step was repeated to record the OD value of each concentration for 8 days.

The micropipette and dropper used are shown in Fig. 4.15. A series of concentration of bacteria were designated as 0.1, 0.2, 0.3, 0.4, 0.5, 0.6, 0.7, 0.8, 0.9 and 1.0 OD. The injection of bacteria broth was gently done using a micropipette to ensure that the concentrations stayed in the artificial cracks. Each specimen was injected with only one concentration of one bacteria. The specimens with artificial cracks

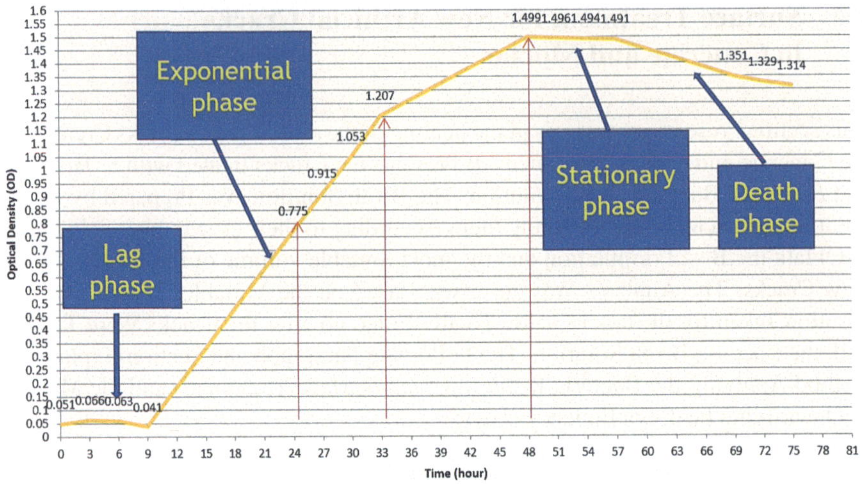


Fig. 4.13 *Sporosarcina pasteurii* bacterial growth curve

**Table 4.1** *S. pasteurii* interval growth rate

| Hours | 0 | 3 | 6 | 9 | **24** | 27 | 30 | **33** |
|---|---|---|---|---|---|---|---|---|
| Optical density (OD) | 0.051 | 0.066 | 0.063 | 0.041 | 0.775 | 0.915 | 1.053 | 1.207 |
| Hours | **48** | 51 | 54 | 57 | 69 | 72 | 75 | |
| Optical density (OD) | 1.499 | 1.496 | 1.494 | 1.491 | 1.351 | 1.329 | 1.314 | |

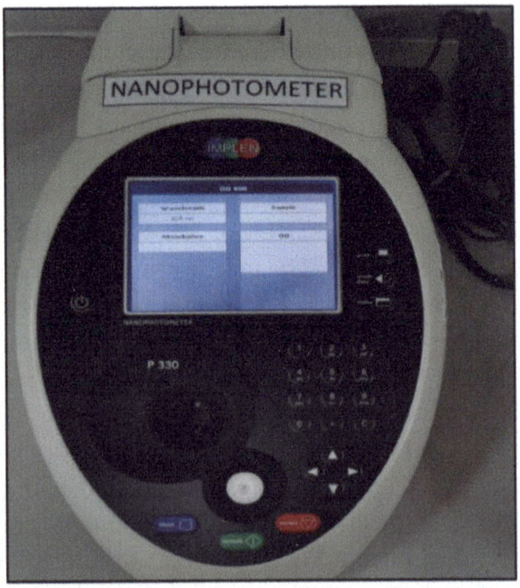


**Fig. 4.14** Nanophotometer to measure the optical density (OD) value

were marked and pointed at three main points. The gap of each mark was 2.5 cm, as shown in Fig. 4.16. An amount of 1 ml of the bacteria in broth solution was dropped into each point area by using a micropipette or dropper.

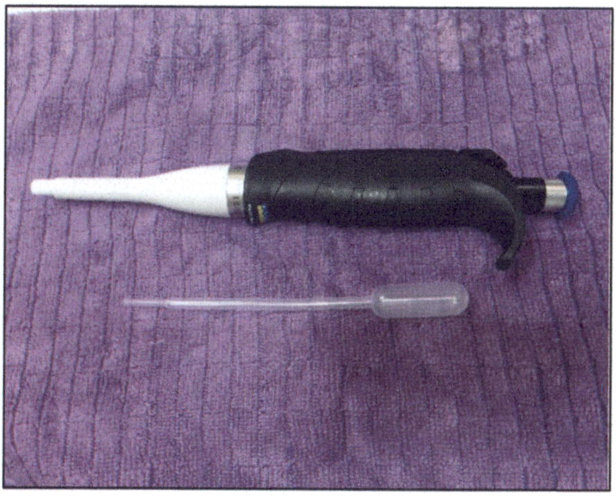

**Fig. 4.15** Image of micropipette and dropper

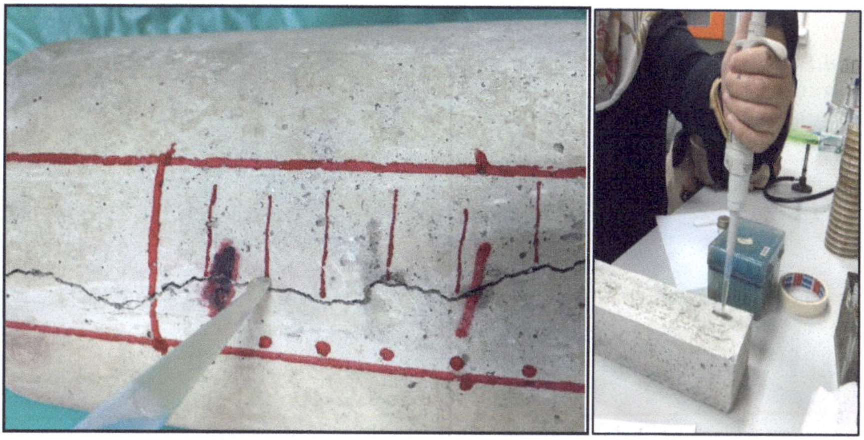

**Fig. 4.16** Bacteria injected in crack of mortar by using micropipette

## 4.5 Quantification of Healing Efficiency (HE)

The evaluation of crack healing was determined by creating different widths of the crack in the specimens. Multiple cylindrical (100 mm × 200 mm) concrete specimens and mortar prisms were used to quantify the crack healing after the bacteria broth solution treatment. A stereo microscope camera Dino Capture 2.0 could visualise the cracks before and after the self-healing treatment (Fig. 4.17).

All cracks were marked as reference points (three marks for each concrete specimen) for injection of bacteria host *S. pasteurii* and other types of bacteria. Results were recorded by visualising the initial ($C_{wi}$) and final images ($C_{wf}$) of the cracks on the concrete specimens. The efficiency of crack healing was detected based on the reduction of the width of the cracks with the presence of calcium carbonate (calcite). Furthermore, the average data from healing were obtained from crack filling and closure. Wang et al. *[33]* suggested the crack healing efficiency (HE) ratio was determined, as in Eq. 4.1.

$$\text{Healing}(100\%) = C_{wi} - C_{wfx}100\% \tag{4.1}$$

where

$C_{wi}$ initial crack width
$C_{wf}$ final crack width

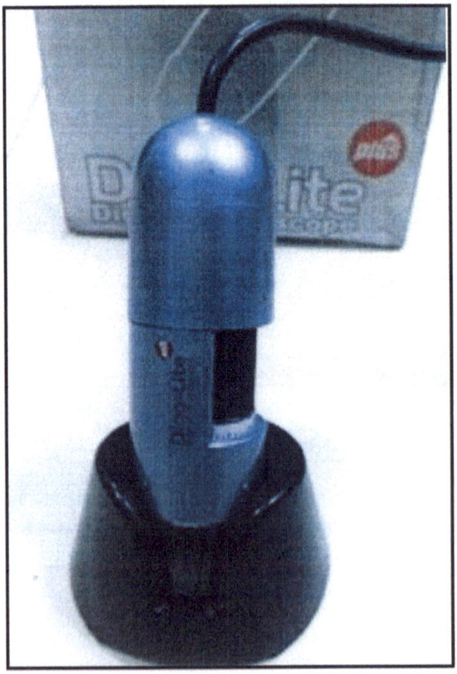

**Fig. 4.17** Stereomicroscope camera (Dino Capture 2.0) to capture crack and healing image

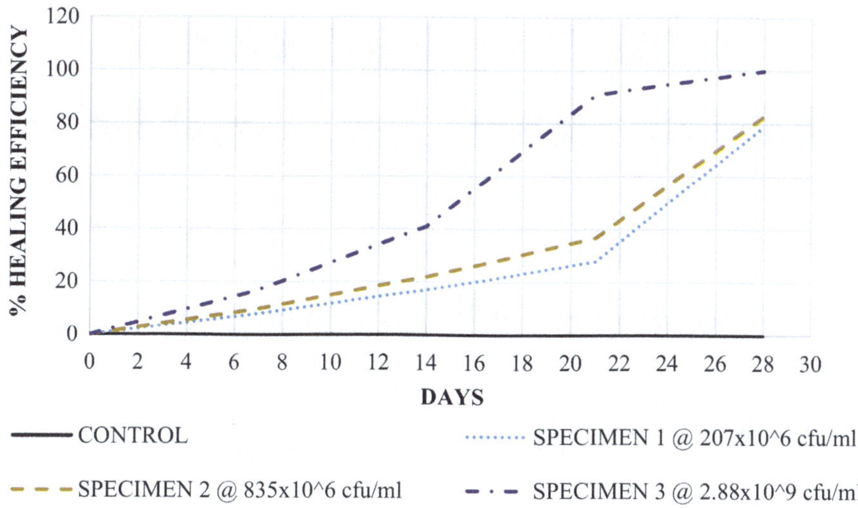


**Fig. 4.18** Healing efficiency (HE) ratio for cracked specimens filled up with solution containing different concentrations of bacteria (broth)

Figure 4.18 shows the healing efficiency (HE) ratio for cracked specimens filled up with broth containing different concentrations of *S. pasteurii* (broth) and the control specimen. It appeared that a higher bacterial concentration of $2.88 \times 10^9$ cfu/ml (48 h) held a greater HE ratio than the cracks treated with other levels of bacteria concentration. In other words, the more increased the bacteria cell concentration used to heal the cracks, the more improved the healing efficiency.

## 4.6 Quantification of Healing on Concrete and Mortar Cracks Versus Cell Concentration

The bacteria broth solution made of *S. pasteurii* bacteria was injected into the first batch of the concrete cracks using a micropipette. Three (3) series of specimens with artificial cracks were used for the application of bacteria taken from three (3) different growth phases, which are at the 24th, 33rd and 48th h after the fermentation process (Fig. 4.13).

The growth at the 48th h reflected the highest population rate, while the 24th h and 33rd h reflected 50% and 75%, respectively before the growth reached the peak level. Four (4) cracked concrete specimens were prepared to be treated, and one (1) control cracked specimen was untreated. The number of live cells in the *S. pasteurii* population was examined at the 24th, 33rd and 48th h. The broth contained bacterial cell concentration, consisting of $207 \times 10^6$ cfu/ml at the 24th h to treat crack in

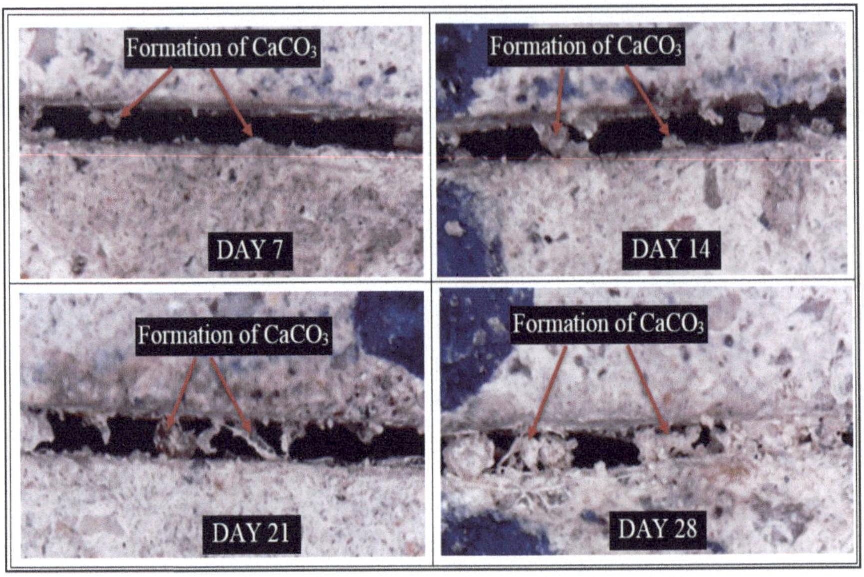


**Fig. 4.19** Crack filling up process captured for specimen 1 after 7, 14, 21 and 28 days of treatment

Specimen 1, 835 × $10^6$ cfu/ml at the 33rd h (Specimen 2) and 2.88 × $10^9$ cfu/ml at the 48th h (Specimen 3).

The images of the healing effect up to 28 days after applying the treatment for each specimen are shown in Figs. 4.19, 4.20 and 4.21 for Specimen 1, Specimen 2 and Specimen 3, respectively. Figure 4.22 shows no healing effect on the cracks created in the control specimens.

Another series of experiments were performed on the cracks of the mortar. A similar type of bacteria, *S. pasteurii,* used to treat the concrete crack specimens was also used on the mortar specimens. The cracks healing process was monitored immediately after the bacteria broth was injected into the cracks. It was also done every day for up to 28 days. Figure 4.22 shows the healing mortar for the artificial cracks after twenty eight (28) days of treatment with *S. pasteurii* broth. The broth containing different cell concentrations was measured using Optical Density (OD) values. A higher OD value reflected higher turbidity, which inferred a higher concentration of bacteria cells in it. A higher concentration of 0.6 and 0.8 OD can produce a higher formation of calcite than 0.2, 0.4 and 1.0 OD, as proven in the healing process. It was found that the one made of 1.0 OD did not contribute to higher precipitation due to the excessive bacteria concentration, which was not an ideal condition for the bacteria to produce calcite at a higher rate. On the other hand, the concentration of 0.1 OD produced a low formation of calcite. The optimum concentration of bacteria was 0.6 OD. A lesser concentration of bacteria was required to produce the amount of calcite formation similar to that of 0.8 OD.

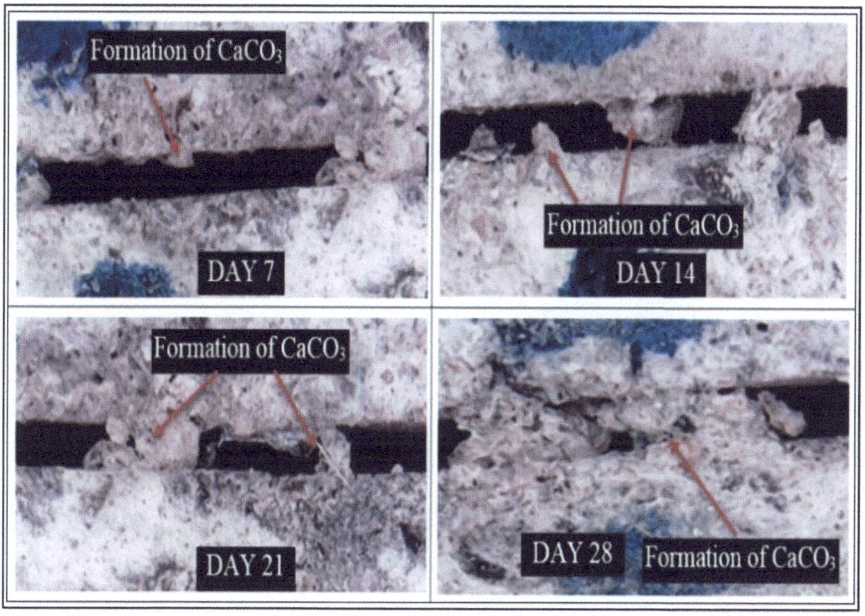


**Fig. 4.20** Crack filling up process captured for specimen 2 after 7, 14, 21 and 28 days of treatment

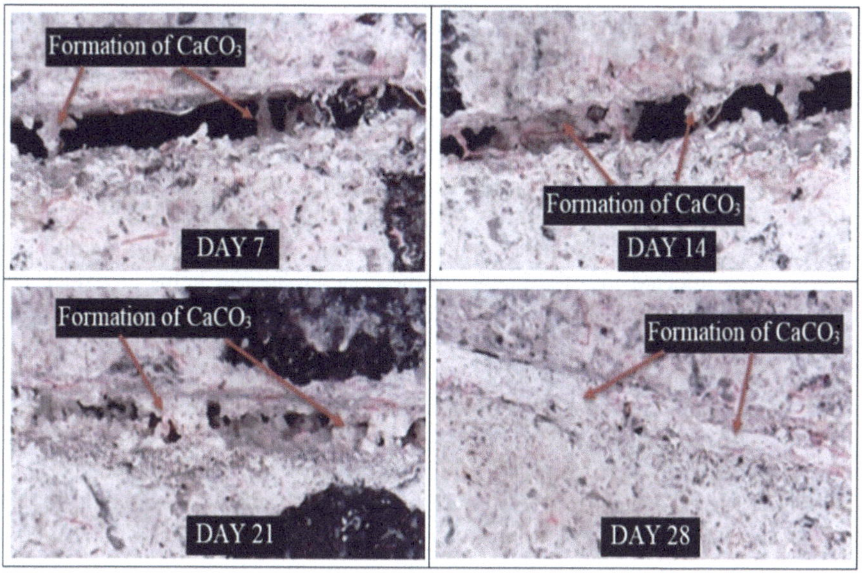


**Fig. 4.21** Crack filling up process captured for specimen 3 after 7, 14, 21 and 28 days of treatment

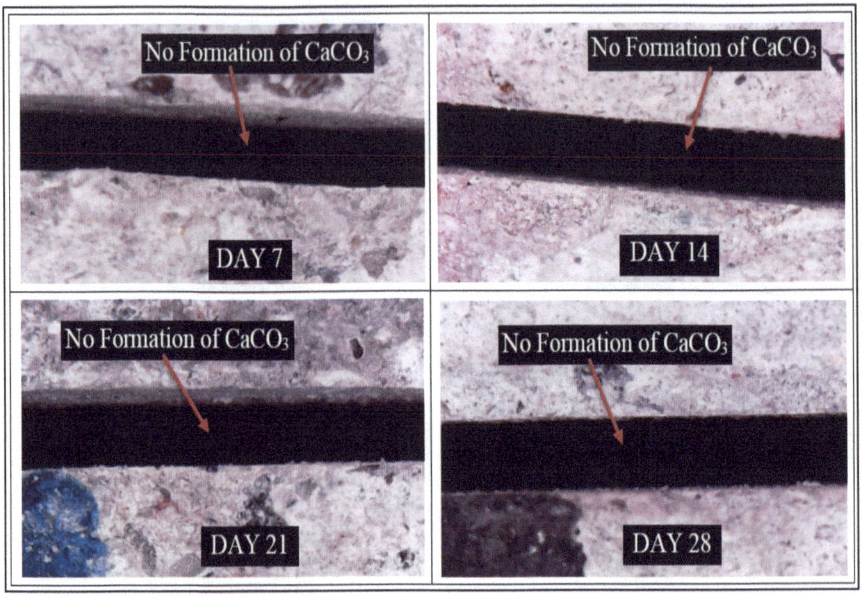


**Fig. 4.22** Crack of control specimen was not filled up after 7, 14, 21 and 28 days of treatment

Bacteria injected into the cracks were activated when they came into contact with water that flowed through the cracks that were dependent on the air humidity level and carried out the conversion of calcium lactate into water-insoluble calcium carbonate that promoted crack-healing. External water is also necessary to activate the bacteria, which convert organic carbon present in the cement-sand matrix into calcium carbonate. The cracks in the concrete could be healed, evidenced by some reduction that occurred in the widths of the concrete or mortar cracks because of the presence of calcium carbonate crystals in the cracks. The sufficient quantity of bacteria being injected into the cracks in the concrete/mortar helps to reduce the crack widths by producing enough formation of calcium carbonate crystals to heal the cracks from within.

It shows that concrete specimens incorporated with 0.2 OD of concentration in *S. pasteruii* bacteria demonstrated the lowest rate of calcite production and could not cover the 0.3 mm width of the crack. Meanwhile, the concrete specimens incorporated with 0.8 OD of *S. pasteruii* bacteria concentration showed the highest rate of calcite production than those treated with broth that recorded 0.2, 0.4 and 0.6 OD of bacteria concentrations. Bacteria could produce calcium carbonate rapidly in 0.8 OD of bacteria concentration due to sufficient urea source to be used by the bacteria *S. pasteruii* as the source of nitrogen, carbon and energy to produce urease enzyme and form carbonate and ammonia. Thus, the bacteria *S. pasteruii* broth with 0.8 OD of concentration can heal the concrete crack at the highest rate.

## 4.7 Examination of Crack Filling Material

The filling materials that healed the cracks were examined using Scanning Electron Microscopy (SEM) and X-Ray Diffraction (XRD). This confirmed that the filling material was calcium carbonate induced from the injected bacteria. The following sub-sections presents how the crack filling materials are confirmed by SEM and XRD.

### *4.7.1 Verification of Filling Crack Materials by SEM*

For further observation, the filling materials of the artificial cracks were subjected to Scanning Electron Microscopy (SEM) and X-Ray Diffraction (XRD) analysis in order to confirm the precipitation that filled up the cracks was either calcite or calcium carbonate derived from the urease activity from the bacteria contained in the broth applied to the cracks. The micrograph image of calcium carbonate-based crystals in the concrete cracks was looked for, as shown in Fig. 4.23. It was expected to prove that calcite crystals were produced by *S. pasteurii* bacteria. The $CaCO_3$ produced microbially was responsible for filling up the cracks in the concrete. The $CaCO_3$ crystals formation should be identical to those reported by Jonkers et al. [16] and Wang et al. [32] whereby the formation in the bacterial activity is derived as in Eq. 4.2:

$$CaC_6H_{10}O_6 + 6O_2 \rightarrow CaCO_3 + 5CO_2 + 5H_2O \quad (4.2)$$

For further observation, all the concrete specimens with the artificial cracks that had been created with aluminium plates and injected with bacteria must be examined via Scanning Electron Microscopy (SEM) to observe any changes in concrete microstructure due to crack-healing and confirm that the calcium carbonate-based crystals (calcite) in the concrete cracks were actually produced by *S. pasteurii* bacteria. The crack healing process was checked after 28 days of treatment. The results showed that the calcite formation resulted from the component of bacteria and a calcium lactate-based healing system. The images of calcite formation for the control specimen, the one that was treated with the broth solution that was made of bacteria, which grew after 33 h and 48 h of fermentation with $835 \times 10^6$ cfu/ml at the 33rd h (Specimen 2) and $2.88 \times 10^9$ cfu/ml at the 48th h *and* are shown in Figs. 4.25, 4.26 and 4.27, respectively. Figure 4.25 shows that no calcite was formed on the non-treated crack.

Figures 4.28 and 4.29 show that the formation of *S. pasteurii* DM 33 induced calcite precipitation on the filling-up material was verified by SEM micrographic. The images of induced calcite by SEM micrographs were taken from the crack filling materials treated with respective bacteria broth solutions. The images of SEM portray that the formation of the $CaCO_3$ crystals was identical to and confirmed with those

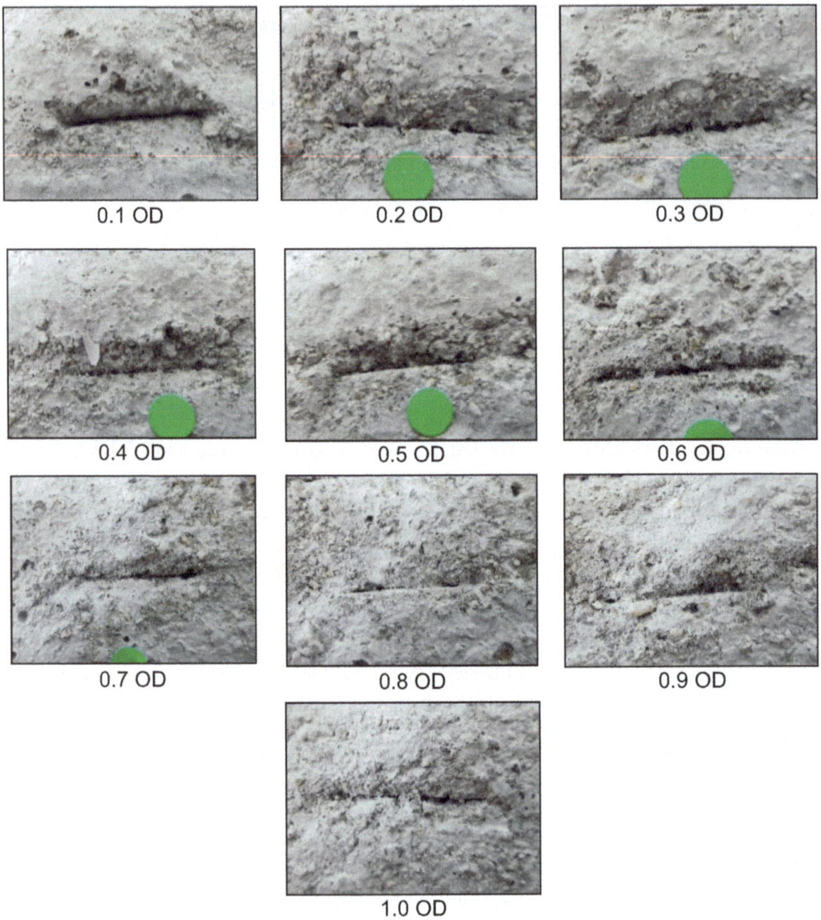


**Fig. 4.23** Images of cracks (0.3 mm width) after being treated with different OD concentrations after 28 days of treatment

reported by Jonkers et al. [16] and Wang et al. [32] which was higher in quantities and had a similar shape (Fig. 4.24).

### *4.7.2 Verification of Filling Crack Materials by X-Ray Diffraction (XRD)*

The morphology and mineralogical compositions of $CaCO_3$ crystals (calcite) were determined by the XRD analysis using PANalytical X'Pert PRO XRD instrument

**Fig. 4.24** Formation of $CaCO_3$ in concrete crack by SEM [17]. Reprinted from Construction and Building Materials, 68, Wasim Khaliq, Muhammad Basit Ehsan, Crack healing in concrete using various bio influenced self-healing techniques, 349–357, Copyright 2015, with permission from Elsevier

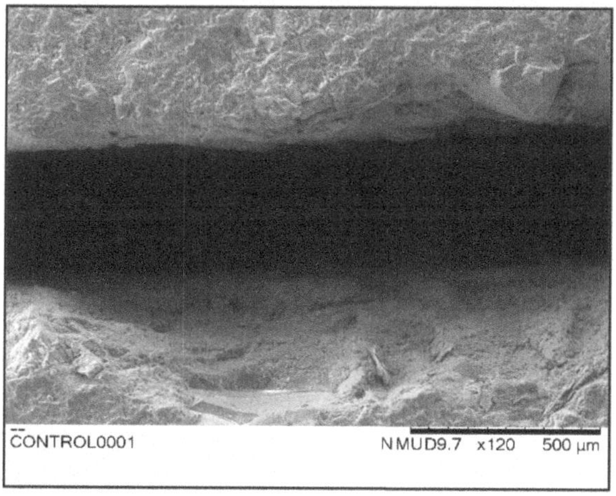


**Fig. 4.25** No formation of calcite was captured by SEM in the crack on the control specimen (without treatment)

(USA) at 40 kV and 30 mA (Fig. 4.30). The XRD analysis was conducted to characterise the crystalline phases of the $CaCO_3$ crystals formed in the cracks. A scraper was used to obtain a sample of the crack filling material from the healed area. The XRD measurement was done at 20 °C and continuously with a fixed divergence slit type at 25 °C using Cu (Copper) as an anode material. For the diffraction pattern of the sample, the raw data of standard $CaCO_3$ was analysed, and the peaks were compared with the peaks obtained from the filling material in the cracks. The results of the X-ray diffractogram show well-defined peaks, as shown in Figs. 4.30, 4.31, 4.32, 4.33, 4.34 and 4.35.

Figure 4.31 shows the standard $CaCO_3$ crystal peak. Figures 4.32 and 4.33 portray the diffractogram pattern examined on the crack filling material that was treated with broth made of *S. pasteurii* DSM 33 and *B. subtilis* ATCC 6633. It is confirmed

**Fig. 4.26** SEM micrographs of the calcite formation in the crack created on the concrete after being treated with broth containing *S. pasteurii* that were taken at the 33rd h of growth

**Fig. 4.27** SEM micrographic of calcite formation in the crack created on the concrete after being treated with broth containing *S. pasteurii* that were taken at the 48th h of growth

**Fig. 4.28** SEM micrographic of the calcite formation derived from *S. Pasteurii* DM 33 was taken in the crack after the treatment

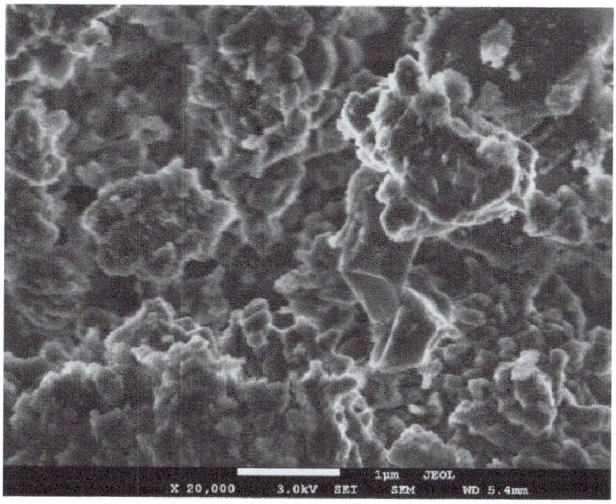


**Fig. 4.29** SEM micrographic of the calcite formation derived from *B. subtilis* ATCC 6633 was taken in the crack after the treatment

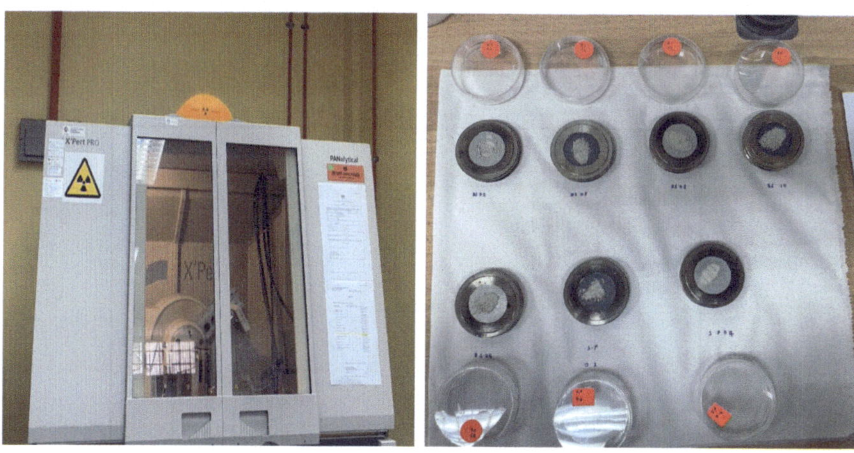

**Fig. 4.30** The XRD instruments and the sample holder of XRD were used to determine the peak diffractogram of the filling crack materials

that calcite ($CaCO_3$) was present through the peaks. A similar peak ratio between calcite (29.3 °C) and vaterite (25 °C) compared with the standard (calcium carbonate ($CaCO_3$)) powder (Fig. 4.31) was observed. The ratio between calcite and vaterite phases was shown to be different for each crack treated with different bacteria. Based on the result obtained, two (2) different isolates, namely *S. pasteurii* DSM 33 and *Bacillus subtilis* ATCC 6633, showed an average peak ratio between calcite (29.3 °C) and vaterite (25 °C).

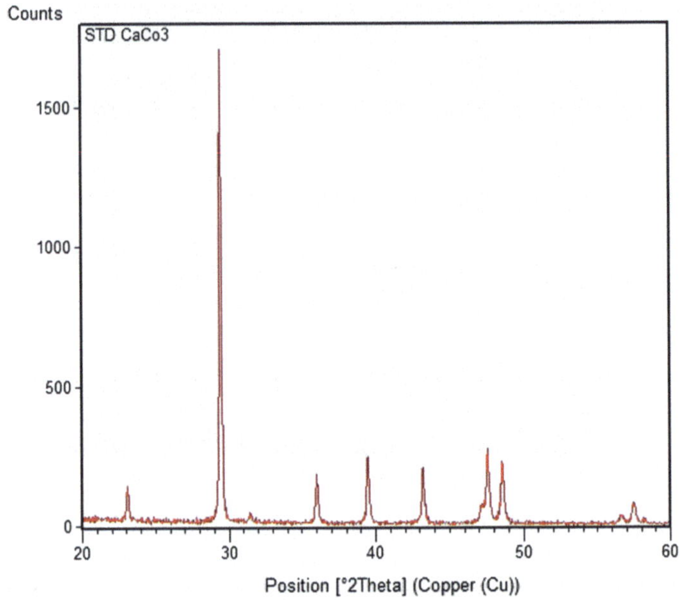


**Fig. 4.31** Calcium carbonate ($CaCO_3$) standard

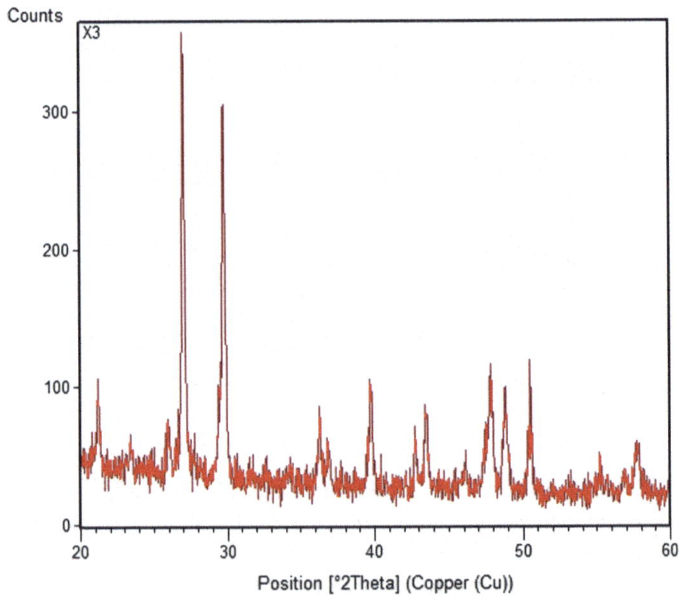


**Fig. 4.32** XRD diffractogram peaks of calcium carbonate crystals formed by *S. pasteurii* DSM 33

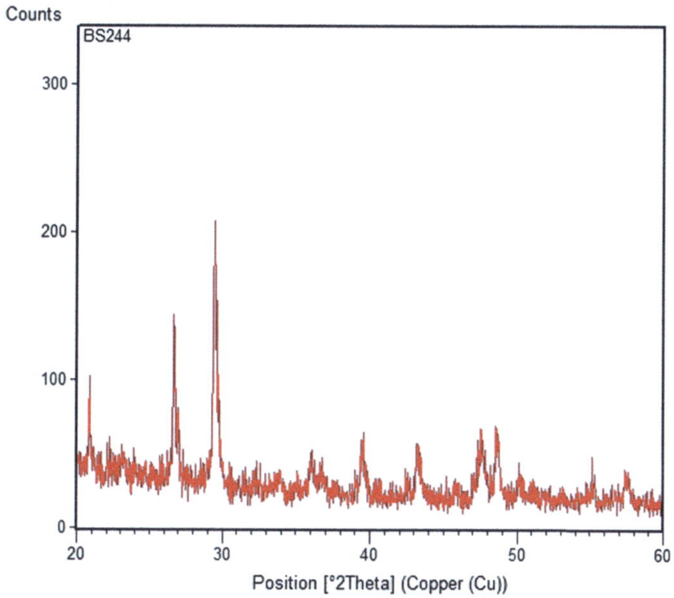


**Fig. 4.33** XRD pattern of calcium carbonate crystals taken from the filling sample that was treated with *B. subtilis* ATCC 6633

Figure 4.32 shows the diffractogram of $CaCO_3$ crystals on the filling sample that was treated with *S. pasteurii* DSM 33. It shows the peak of the sample was the highest of the peak ratio at 26 °C and 29 °C, while the highest peak ratio of those treated with *B. subtilis* ATCC 6633 was at 29°, as illustrated in Fig. 4.33. The XRD patterns showed that each of the bacteria isolates was able to produce calcite and vaterite phases. These two phases were the main elements of calcium carbonate. Besides that, the peak ratio showed a value similar to the calcium carbonate standard. The filling crack material that was derived from the one treated with *S. pasteurii* DSM 33 broth showed a higher peak of calcite and vaterite in the intensity than the one that was treated with *B. subtilis* ATCC 6633 isolate. The crystalline calcite, a crucial component in the healing process, is developed in the presence of carbon dioxide and leached out of the fractures to become hard [11].

With the presence of calcite, the XRD analysis confirmed that crack-healing had occurred, and this confirmation was corroborated by the micrographs obtained from the SEM examination. Both XRD analysis and the SEM examination also showed that calcite was the most common mineral being formed when bacteria were present.

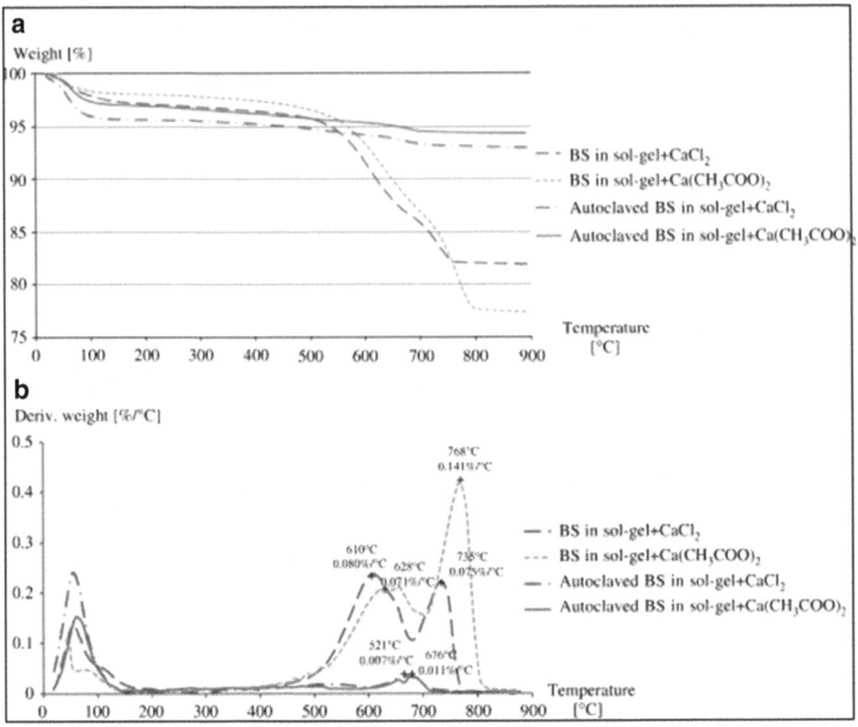


**Fig. 4.34** TGA results for different crack repair materials containing active or auto-claved bacteria, **a** decrease in weight versus temperature, and **b** derivative of weight loss versus increase in temperature [29]. Reprinted from Cement and Concrete Research, 40, Kim Van Tittelboom, Nele De Belie, Willem De Muynck, Willy Verstraete, Use of bacteria to repair cracks in concrete, 157–166, Copyright 2009, with permission from Elsevier

## 4.8 Confirmation of Calcite ($CaCO_3$) Formation of by Thermogravimetric Analysis (TGA)

Thermogravimetric analysis (TGA) reported in the literature as one of the methods to detect the formation of calcite in the observed material through mass decomposition. This approach selected because different materials have various thermal stability [10] in which the proper material identification are distinguishable. Interpreting the TGA curve is a crucial aspect of extracting meaningful information about the thermal behavior of a sample [4]. There are two (2) ways of interpreting data from TGA curves; (1) the percentage of weight loss plotted against temperature; and (2) the derivative of the weight loss plotted against the temperature [29].

Table 4.2 shows the temperature range of TGA analysis as obtained from Wang et al. [34]. Each phase (including the hydrates) is characterised by its own temperature range of decomposition and by a specific mass loss; as an example, for portlandite, it is a loss of water, and for calcite, it is a loss of $CO_2$. The lowest temperature of

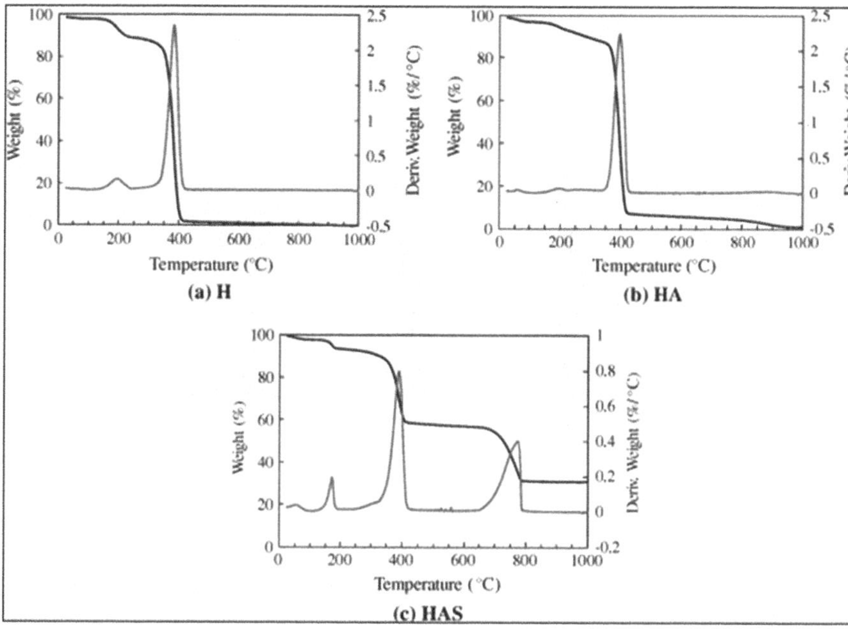


**Fig. 4.35** TGA graphs of samples from various types of hydrogel pieces soaked in water (the black and grey lines are the weight (percentage) changes, and the derivative of the weight (percentage) changes as a function of temperature, respectively) [35]. Reprinted from Construction and Building Materials, 68, J. Y. Wang, D. Snoeck, S. Van Vlierberghe, W. Verstraete, N. De Belie, Application of hydrogel encapsulated carbonate precipitating bacteria for approaching a realistic self-healing in concrete, 110–119, Copyright 2014, with permission from Elsevier

the dissociation range is determined by the characteristics of the equipment and the heating rate. The maximum temperature of the dissociation range is a function of the quantity of the studied phase. The temperature ranges are clearly defined by the edges of the characteristic peak of the Differential Thermogravimetric (DTG) curve.

Different temperature range between 500 °C and 800 °C were used by Wang et al. [34], Van Tittleboom et al. (2010), and Kong et al. [19] on calcium carbonate formation. Control specimens were prepared to ensure the formed calcium carbonate was mainly due to the presence of bacteria which adhered to the carbonate precipitation. The control samples represent as non-bacterial specimens which comprised of specimens with free bacteria, autoclaved bacteria and also a pure form of encapsulation.

Research conducted by Van Tittelboom [29] emphasise on different results obtained by using *Bacillus sphaericus* (BS) in active and autoclaved (dead) bacterial cells. Both types of bacteria encapsulated in silica gel with biomineral precursor compound which consists of calcium chloride ($CaCl_2$) and calcium acetate ($Ca(CH_3COO)_2$). The TGA analysis data used as an interpretation of bacteria influence towards calcite formation shown in Fig. 4.34. Figure 4.34a shows that at about

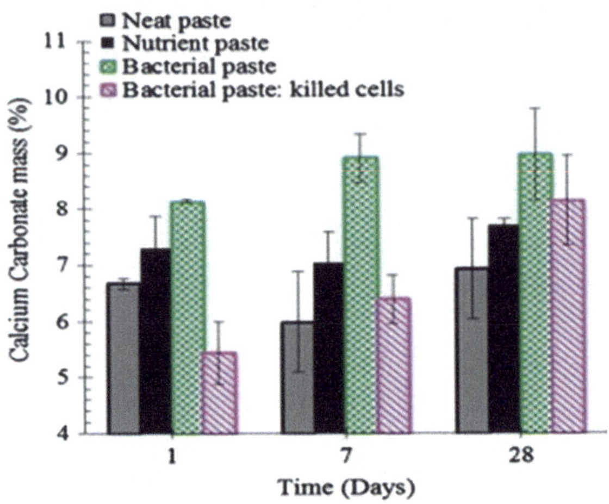


**Fig. 4.36** Calcium carbonate content in neat paste, nutrient paste, bacterial paste, and bacterial paste with killed cells at 1, 7, and 28 days [7]. Reprinted from Cement and Concrete Research, 67, Zeynep Basaran Bundur, Mary Jo Kirisits, Raissa Douglas Ferron, Biomineralized cement-based materials: Impact of inoculating vegetative bacterial cells on hydration and strength, 237–245, Copyright 2014, with permission from Elsevier

**Table 4.2** Temperature range of thermogravimetric analysis (TGA) [34]

| Hydration progress | Temperature range (°C) | Description |
|---|---|---|
| Dehydration | 105–1000 | Continuous reduction in weight due to loss of the chemically combined hydrate water from the gel products which occur throughout the temperature range |
| Degradation | 200–450 | Degradation of the alginate hydrogel and Weight loss due to hydration |
| Dehydroxylation | 450–550 | Dehydroxylation of calcium hydroxide that leads to weight loss due to the water steaming |
| Decomposition | 600–700 | The decomposition of calcium carbonate from the reaction between calcium hydroxide and atmospheric carbon dioxide |

the temperature of 100 °C, the evaporation of water occurs leading to weight decrease. Further continuous of weight decrease detected at a temperature of 500 and 800 °C. These may be due to the decomposition of calcium carbonate ($CaCO_3$). The samples treated with active and autoclaved bacteria has a distinguishable weight loss.

The weight loss is rather small for treatment with autoclaved bacteria with 1.69% and 1.34%, for samples with $CaCl_2$ and $Ca(CH_3COO)_2$ as calcium sources, respectively. However, a sharp decrease in weight observed for treatment with active bacteria at 13.64% and 18.85%, for samples with $CaCl_2$ and $Ca(CH_3COO)_2$ respectively. These results provide evidence that $CaCO_3$ crystals are formed particularly in the case of using active bacteria.

Meanwhile, Fig. 4.34b shows the derivative in weight loss versus temperature to indicate the points at which the weight loss is most apparent. Only a small amount of $CaCO_3$ is decomposed at the temperature of 521 °C with $CaCl_2$ as a calcium source and 676 °C with $Ca(CH_3COO)_2$ as a calcium source for the repair materials containing autoclaved bacteria. This chemical precipitation of $CaCO_3$ may attribute to the weight decrease. The observed peaks are more distinct with two different peaks are detected for each material for the repair materials containing active bacteria and as can be seen in the Fig. 4.34b.

TGA results demonstrated the presence of $CaCO_3$ precipitation in/on the bio-hydrogels from the study conducted by Wang et al. [35] as illustrated in Fig. 4.35. $CaCO_3$ decomposes into CaO and $CO_2$ in the temperature range of 650–800 °C and this causes weight loss [22]. As shown in Fig. 4.35a, b, the samples from H and HA had no weight loss occurring around 650–800 °C which indicated that no $CaCO_3$ formed in these samples. While for the hydrogels incorporated with bacterial spores (HAS), distinct weight loss in the temperature range of 650–800 °C was observed (Fig. 4.35c). The weight loss percentage was about 26%. It also noticed that in all graphs (**a-c**), there was a sharp weight loss in the temperature range of 340–400 °C, which was due to the degradation of the hydrogel itself. Obviously, due to the presence of $CaCO_3$, the weight loss percentage from hydrogel degradation in the bio-hydrogels was smaller than in the ones without bacteria. Another approach to data presented by Bundur et al. [7] using a bar chart to represent the calcium carbonate decomposition mass as shown in Fig. 4.36.

Four (4) types of paste prepared namely neat paste (prepared with distilled water and cement), nutrient paste (prepared with fresh Urea-Yeast Extract (UYE) medium and cement, bacterial paste (made with vegetative cells of *Sporocarsina pasteurii* grown in UYE medium, cement and the vegetative cells of *Sporocarsina pasteurii* in the form of bacterial paste were incorporated into neat paste, nutrient paste, bacterial paste with killed cells of *Sporocarsina pasteurii.* The temperature used to detect calcium carbonate formation was 700 and 900 °C. Based on Fig. 4.35, the calcium carbonate content was noticeably higher in the bacterial paste as compared to the nutrient paste and neat paste. The addition of UYE medium increased the calcium carbonate content in the cement paste (nutrient paste), the addition of vegetative bacterial culture increase pronouncedly the decomposition mass of calcium carbonate (bacterial paste). The vegetative cells of *Sporocarsina pasteurii* cells increased the average amount of calcium carbonate by 22% to 42% of neat paste at 1 and 7 days of curing, respectively. The impact of fresh UYE medium (nutrient paste) on calcium carbonate content was substantially lower than the impact of the bacterial culture at 1 and 7 days of curing. Thus, the results demonstrate that calcium carbonate content increased with the inoculation of a bacterial culture consisting of vegetative *Sporocarsina pasteurii* cells in UYE medium. At day 1, the incorporation of killed cells in UYE medium in the bacterial paste resulted from a reduction in calcium carbonate content as compared to neat paste. However, at day 7, the calcium carbonate content of the sample containing the killed cells was similar to that of neat paste. Also, it should note that autoclaving the medium to kill the cells likely drove $CO_2$ from solution, though equilibrium with atmospheric $CO_2$ could be re-established over time. Thus, an autoclaving method could have impacted $CaCO_3$ precipitation negatively in the bacterial paste with killed cells, in such data shown that $CaCO_3$ precipitation initially was less than that in the bacterial paste. However, on 28 days of curing, the calcium carbonate content of the bacterial paste with killed cells was on par with the sample containing bacterial paste with vegetative cells. Overall, these results revealed that the presence of vegetative cells in concrete is capable of maximising the formation of calcium carbonate precipitation at an early age (up to 7 days).

The TGA instrument is the easiest and most widely used method to detect the formation of calcium carbonate that presence in material [30]. However, different temperature ranges for calcium carbonate formation were proposed in literature [1, 2, 7, 29, 31] (Jamilu 2015, Stuckrath et al. 2014). Archer Jr et al. [2] and Villain et al. [30] stated that the particle size and type of instrument with different pressure rates would influence the temperature range of calcium carbonate formation. In a study conducted by Archer Jr et al. [2] on the determination of calcium carbonate peak temperature. Calcium carbonate with size less than 150 μm, has a decomposition peak temperature at 719.0 to 807.0 °C. Meanwhile, in a larger size of calcium carbonate, the temperature peak of decomposition also higher. The calcium carbonate with the size of 3 mm, the temperature peak were 933.7 to 971.7 °C. The results revealed that the bigger particle size, the higher temperature peak of material decomposition behaviour.

In summary, the formation of calcite as a result of Microbially Induced Calcite Precipitation (MICP) can be verified through various analytical methods. Visual inspection, utilising high-resolution cameras and light microscopy, provides preliminary evidence of surface crack closure. For greater depth of observation and to assess microstructural details, SEM is more appropriate, enabling the identification of different calcite morphologies. Among these, the rhombohedral shape is recognised as the most stable and characteristic form of calcite. To further confirm the presence of calcite, TGA and XRD could be employed. TGA detects calcite formation through thermal decomposition profiles, as each material exhibits distinct thermal stability, allowing for clear differentiation of components. Meanwhile, XRD is a powerful analytical technique used to determine the crystalline structure, morphology, and mineralogical composition of calcium carbonate ($CaCO_3$) crystals, particularly in the form of calcite. In the context of crack healing in concrete, XRD is employed to identify and characterise the crystalline phases of $CaCO_3$ formed within the cracks. The diffraction patterns obtained provide evidence of the presence and type of $CaCO_3$ polymorphs—such as calcite, aragonite, or vaterite—based on their unique crystallographic signatures. This information is crucial for confirming the effectiveness of MICP and understanding the stability and durability of the precipitated minerals in self-healing concrete systems.

## References


1. Abo-El-Enein, S.A., Ali, A.H., Talkhan, F.N., Abdel-Gawwad, H.A.: Utilization of microbial induced calcite precipitation for sand consolidation and mortar crack remediation. HBRC J. **8**(3), 185–192 (2012). https://doi.org/10.1016/j.hbrcj.2013.02.001
2. Archer Jr., P.D., Ming, D.W., Sutter, B.: The effects of instrument parameters and sample properties on thermal decomposition: interpreting thermal analysis data from mars. Planet. Sci. **2**(1), 1–21 (2013)
3. Aishwarya, T., Juneja, A.: Efficiency and morphology of calcium carbonate precipitate induced by urease enzymes. Geotech. Geol. Eng. **42**(2), 1153–1171 (2024). https://doi.org/10.1007/s10706-023-02610-6
4. Akash, M.S.H., Rehman, K.: Comprehensive insights into thermo gravimetric analysis. In: Essentials of Pharmaceutical Analysis, pp. 899–929. Springer (2025)
5. Anbu, P., Kang, C.-H., Shin, Y.-J., So, J.-S. (2016). Formations of calcium carbonate minerals by bacteria and its multiple applications. Springerplus 5(1), 1–26. doi:https://doi.org/10.1186/s40064-016-1869-2
6. Andalib, R., Muhd Zaimi, A.M., Mohd Warid, H., Mohanadoss, P., Ali, K., Jahangir, M., Han-Seung, L.: Optimum concentration of *Bacillus megaterium* for strengthening structural concrete. Constr. Build. Mater. **118**, 180–193 (2016). https://doi.org/10.1016/j.conbuildmat.2016.04.142
7. Bundur, Z.B., Kirisits, M.J., Ferron, R.D.: Biomineralized cement-based materials: impact of inoculating vegetative bacterial cells on hydration and strength. Cem. Concr. Res. **67**, 237–245 (2015). https://doi.org/10.1016/j.cemconres.2014.10.002
8. Chen, Y.-Q., Wang, S.-Q., Tong, X.-Y., Kang, X.: Crystal transformation and self-assembly theory of microbially induced calcium carbonate precipitation. Appl. Microbiol. Biotechnol. **106**(9), 3555–3569 (2022)
9. da Silva, F.B., De Belie, N., Boon, N., Verstraete, W.: Production of non- axenic ureolytic spores for self-healing concrete applications. Constr. Build. Mater. **93**, 1034–1041 (2015). https://doi.org/10.1016/j.conbuildmat.2015.05.049
10. Frost, R., Hales, M., Martens, W.: Thermogravimetric analysis of selected group (ii) carbonate minerals—implication for the geosequestration of green-house gases. J. Therm. Anal. Calorim. **95**(3), 999–1005 (2008)
11. Ganesh, C.A., Muthukannan, M., Malathy R, & Babu, R.C. (2019) An experimental study on effects of bacterial strain combination in fibre concrete and self-healing efficiency. KSCE J. Civ. Eng. 1–10. doi:https://doi.org/10.1007/s12205-019-1661-2
12. Gorospe, C.M., Han, S.H., Kim, S.G., Park, J.Y., Kang, C.H., Jeong, J.H., So, J.S.: Effects of different calcium salts on calcium carbonate crystal formation by Sporosarcina pasteurii KCTC 3558. Biotechnol. Bioprocess Eng. **18**(5), 903–908 (2013)
13. Hong, S., Ren, H., Hou, D., Dong, B., Fan, S.: Visual and quantitative evaluation of crack bonding rate of self-healing mortars based on volumetric strain analysis. J. Clean. Prod. **450**, 141928 (2024). https://doi.org/10.1016/j.jclepro.2024.141928
14. Jiang, L., Wu, M., Du, F., Chen, D., Xiao, L., Chen, W., Du, W., & Ding, Q. (2024). State-of-the-art review of microcapsule self-repairing concrete: principles, applications, test methods, prospects. Polymers 16(22), 3165
15. Jonkers, H., Schlangen, E.: Crack repair by concrete-immobilized bacteria. In: Proceedings of the First International Conference on Self Healing Materials, pp. 1–7 (2007)
16. Jonkers, H., Thijssen, A., Muyzer, G., Copuroglu, O., Schlangen, E.: Application of bacteria as self-healing agent for the development of sustainable concrete. Ecol. Eng. **36**(2), 230–235 (2010). https://doi.org/10.1016/j.ecoleng.2008.12.036
17. Khaliq, W., Ehsan, M.B. (2016) Crack healing in concrete using various bio influenced self-healing techniques. Constr. Build. Mater. **102**, Part 1, 349–357. https://doi.org/10.1016/j.conbuildmat.2015.11.006

18. Krishnapriya, S., Venkatesh Babu, D.L., Prince Arulraj, G.: Isolation and identification of bacteria to improve the strength of concrete. Microbiol. Res. **174**, 48–55 (2015). https://doi.org/10.1016/j.micres.2015.03.009
19. Kong, Q.-S., Wang, B.-B., Ji, Q., Xia, Y.-Z., Guo, Z.-X., Yu, J.: Thermal degradation and flame retardancy of calcium alginate fibers. Chin. J. Polym. Sci. **27**(06), 807–812 (2009)
20. Merchant, F., Castleman, K. (eds.): Microscope Image Processing. Academic (2022)
21. Okwadha, G.D.O., Li, J.: Optimum conditions for microbial carbonate precipitation. Chemosphere. **81**(9), 1143–1148 (2010). https://doi.org/10.1016/j.chemosphere.2010.09.066
22. Oniyama, E., Wahlbeck, P.: Application of transpiration theory to TGA data: calcium carbonate and zinc chloride. Thermochim. Acta. **250**(1), 41–53 (1995)
23. Qian, C., Ren, L., Xue, B., Cao, T.: Biomineralization on cement-based ma- terials consuming $CO_2$ from atmosphere. Constr. Build. Mater. **106**, 126–132 (2016)
24. Qian, C., Jianyun, W., Ruixing, W., Liang, C.: Corrosion protection of cement-based building materials by surface deposition of $CaCO_3$ by bacillus pasteurii. Mater. Sci. Eng. C. **29**(4), 1273–1280 (2009)
25. Qiu, J., Tng, D.Q.S., Yang, E.-H.: Surface treatment of recycled concrete aggregates through microbial carbonate precipitation. Constr. Build. Mater. **57**, 144–150 (2014). https://doi.org/10.1016/j.conbuildmat.2014.01.085
26. Rose, I. (2012) Light microscope vs electron microscope. http://www.ivyroses.com/Biology/Techniques/light-microscope-vs-electron-microscope.php
27. Tobler, D.J., Cuthbert, M.O., Greswell, R.B., Riley, M.S., Renshaw, J.C., Handley-Sidhu, S., Phoenix, V.R.: Comparison of rates of ureolysis between *Sporosarcina pasteurii* and an indigenous groundwater community under conditions required to precipitate large volumes of calcite. Geochim. Cosmochim. Acta. **75**(11), 3290–3301 (2011). https://doi.org/10.1016/j.gca.2011.03.023
28. Tong, H., Ma, W., Wang, L., Wan, P., Hu, J., & Cao, L. (2004). Control over the crystal phase, shape, size and aggregation of calcium carbonate via a l-aspartic acid inducing process. Biomaterials 25(17), 3923–3929. https://doi.org/10.1016/j.biomaterials.2003.10.038
29. Van Tittelboom, K., De Belie, N., De Muynck, W., Verstraete, W.: Use of bacteria to repair cracks in concrete. Cem. Concr. Res. **40**(1), 157–166 (2010). https://doi.org/10.1016/j.cemconres.2009.08.025
30. Villain, G., Thiery, M., Platret, G.: Measurement methods of carbonation profiles in concrete: thermogravimetry, chemical analysis and gammadensimetry. Cem. Concr. Res. **37**(8), 1182–1192 (2007)
31. Wan, K., Xu, Q., Wang, Y., Pan, G.: 3D spatial distribution of the calcium carbonate caused by carbonation of cement paste. Cem. Concr. Compos. **45**, 255–263 (2014)
32. Wang, J., Van Tittelboom, K., De Belie, N., Verstraete, W.: Use of silica gel or polyurethane immobilized bacteria for self-healing concrete. Constr. Build. Mater. **26**(1), 532–540 (2012). https://doi.org/10.1016/j.conbuildmat.2011.06.054
33. Wang, J., Dewanckele, J., Cnudde, V., Van Vlierberghe, S., Verstraete, W., De Belie, N.: X-ray computed tomography proof of bacterial-based self-healing in concrete. Cem. Concr. Compos. **53**, 289–304 (2014b)
34. Wang, J., Mignon, A., Snoeck, D., Wiktor, V., Van Vliergerghe, S., Boon, N., De Belie, N.: Application of modified-alginate encapsulated carbonate producing bacteria in concrete: a promising strategy for crack self-healing. Front. Microbiol. **6**(1088), 1–14 (2015). https://doi.org/10.3389/fmicb.2015.01088
35. Wang, J., Snoeck, D., Van Vlierberghe, S., Verstraete, W., De Belie, N.: Application of hydrogel encapsulated carbonate precipitating bacteria for approaching a realistic self-healing in concrete. Constr. Build. Mater. **68**, 110–119 (2014a)
36. Wang, W., Zhu, M., Zou, C., Zhang, T., Han, J., Wang, Z.: Biosurfactant type and introduction order impact the morphology of biological $CaCO_3$ during the MICP process. World J. Microbiol. Biotechnol. **41**(7), 205 (2025). https://doi.org/10.1007/s11274-025-04445-0
37. Wiktor, V., Jonkers, H.M.: Quantification of crack-healing in novel bacteria-based self-healing concrete. Cem. Concr. Compos. **33**(7), 763–770 (2011). https://doi.org/10.1016/j.cemconcomp.2011.03.012

38. Zhang, J., Xiao, Y., Liu, H., & Chu, J. (2025). Role of bacteria on bio-induced calcium carbonate formation: insights from droplet microfluidic experiments. Géotechnique 75(6), 787–799
39. Zhang, S., Ding, J., Wang, S., Li, C.: Evaluation of calcium carbonate production and cementitious characteristics of enzymatically induced carbonate precipitation during environmental adjustment. Constr. Build. Mater. **449**, 138415 (2024). https://doi.org/10.1016/j.conbuildmat.2024.138415

# Chapter 5
# Suitability of Alginated Hydrogel Encapsulated *Geobacillus stearothermophilus* as Self-Healing Agent in Concrete

## 5.1 Introduction

This chapter provides an examination of the physical characteristics of encapsulated of *Geobacillus stearothermophilus*, focusing on its roles and performance as a self healing agent. This is an intention of the authors to demonstrate the survival ability of *Geobacillus stearothermophilus* when encapsulated into alginate hydrogel and its potential to precipitate calcium carbonate ($CaCO_3$) when incorporated into the cementitious material. The success of using *Geobacillus stearothermophilus* will provide the basis for better understanding of using microbes as a self-healing agent in cementitious material, which eventually increase the fundamental and substances to the pool of existing knowledge for future research. The *Geobacillus stearothermophilus* is another bacteria from genus *Bacillus* was selected as a highlight in this study. The *Geobacillus stearothermophilus* can be found in the geothermally heated region or high-temperature ambient such as hot spring, and it is a highly resistant bacteria due to its capability thriving high temperature (30–75 °C) [1, 7, 11, 14]. The *Geobacillus stearothermophilus* capable to produce endospore by permitted harsh situation [5, 9, 10, 17]. The morphology and resistance properties of *Geobacillus* enable them to be mobilised in the atmosphere and transported within the long distance.

*Geobacillus stearothermophilus* were encapsulated into alginate-hydrogel prior incorporation into the mortar or simulated environment. The following sections will present the urease activity, viability, swelling and water retention properties of the bacterial *Geobacillus stearothermophilus* cell encapsulated in alginate hydrogel. Consequently, the precipitation of calcium carbonate ($CaCO_3$) of alginate-encapsulated with *Geobacillus stearothermophilus* (AE-GS) mortar was confirmed using Thermogravimetric Analysis (TGA).

 
R. M. A. M. Azam and D. H. M. Saman, *Self-Healing Concrete*, Engineering Materials,
https://doi.org/10.1007/978-981-95-7553-4_5

## 5.2 Swelling Capacity and Retention Properties of the Alginate-Encapsulated *Geobacillus stearothermophilus*

The swelling capacity and retention properties of alginate-encapsulated *Geobacilus stearothermophilus* incorporating in cement slurry (C) and distilled water (DW) before incorporating in the mortar are explained in the following sub-sections.

### 5.2.1 Swelling Capacity

Figure 5.1 depicts the swelling capacity for the alginate-encapsulated without *Geobacillus stearothermophilus* (AE) and with *Geobacillus stearothermophilus* (AE-GS) in cement slurry (C) and the distilled water (DW). The values of the swelling capacity obtained from three (3) trials, namely Trial 1 to Trial 3. Based on Fig. 5.1, the swelling capacity for the alginate-encapsulated without *Geobacillus stearothermophilus* in cement slurry (AE-C) is about 5.3 g/g, and with *Geobacillus stearothermophilus* in cement slurry (AE-GSC) is about 6.7 g/g. On the other hand, the swelling capacity is about 7.2 g/g for the alginate-encapsulated without *Geobacillus stearothermophilus* immersed in distilled water (AE-DW) while 8.9 g/g for those with *Geobacillus stearothermophilus* immersed in distilled water (AE-GSDW). Table 5.1 tabulates swelling capacity results of the alginate-encapsulated with *Geobacillus stearothermophilus* (AE-GS) and without the bacteria (AE) with other types of the hydrogel.

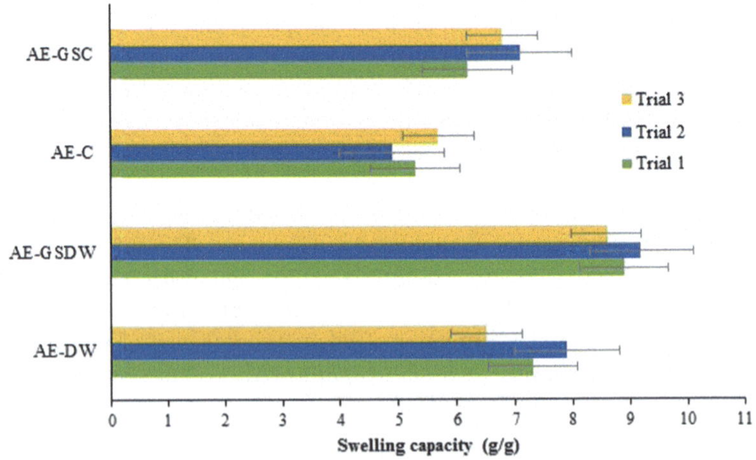


**Fig. 5.1** Swelling capacity of alginate hydrogel immersed in cement slurry and distilled water

**Table 5.1** Swelling capacity of alginate hydrogel tested at 28 days in the present study with those investigated by previous researchers

| References | Types of hydrogel | Condition | Swelling capacity (g/g) | % Reduction of swelling capacity in mimicked mortar environment with respect to water |
|---|---|---|---|---|
| Raden Maizatul Aimi [8] | Alginate hydrogel | Distilled water | 8.9 | – |
| | | Cement slurry | 6.7 | 24.7 |
| Wang et al. [12] | Modified Alginate hydrogel | Dimineralised water | 63.1 | – |
| | | Cement filtrate | 37.0 | 41.4 |
| Wang et al. [13] | Pure hydrogel | De-ionised water | 23.0* | – |
| | | Cement filtrate | 19.0* | 17.4 |

* Interpolated and derived from graph

Wang et al. [13] performed a swelling test on hydrogel loaded with spores and nutrient embedded prior immersion in de-ionised water and filtered cement slurry. The hydrogel immersed in de-ionised water swelling at 23.0 g/g. Meanwhile, the hydrogel expands at 19.0 g/g in filtered cement slurry.

In a study conducted by Wang et al. [12], modified alginate was immersed in both demineralised water and cement filtrate in the absence of bacteria. The results showed that the swelling capacity was higher when the modified alginate was immersed in demineralised water (63.1 g/g) compared to cement filtrate (37.0 g/g). The reduction in swelling capacity was calculated using the hydrogel soaked in demineralised water as the control. As shown in Table 5.1, the highest reduction in swelling capacity 41.4% was recorded for the hydrogel studied by J. Wang et al. [12], where the swelling test was conducted without the presence of bacteria.

The results demonstrated a higher swelling capacity when the hydrogel was immersed in water compared to cement filtrate. This is because the calcium ions present in the cement promote increased cross-linking within the hydrogel matrix, initiating and enhancing the gelation process. As a result, the hydrogel becomes more stable, making it better suited for encapsulating bacteria to facilitate self-healing [6, 12].

For both immersions (cement slurry (C) and distilled water (DW)) conditions, the alginate encapsulation loaded with bacteria has a higher water uptake as compared to those of non-loaded bacteria. The results showed that the *Geobacillus stearothermophilus* has a direct relation on water absorption through the alginate membrane. The addition of calcium ion in cement slurry plays an important role increasing a gelling effect of alginate hydrogel. Swelling capacity is an important parameter in hydrogel as the capability of the hydrogel to swell rather than dissolved when water enters it [18]. The swelling capacity also determines increasing rate of the hydrogel by the presence of water upon formation of crack as excessive of swelling capacity would cause the hydrogel to break quickly with the sudden release of the bacteria [4] along with nutrient embedded inside the hydrogel. The AE-GS swelling rate obtained was comparable of those reported in the literature. Therefore, the swelling properties of alginate-encapsulated without bacteria (AE-C and AE-DW) and with bacteria (AE-GSC and AE-GSDW) obtained is in an acceptable state and able to use as the self-healing agent.

### 5.2.2 Water Retention Properties

Figure 5.2 depicts the percentage of weight changes obtained by water retention test conducted on by alginate-encapsulated gel without *Geobacillus stearothermophilus* (AE) and alginate-encapsulated with *Geobacillus stearothermophilus* (AE-GS). AE-GS capable of retaining 65% of the absorbed water for first 12 h, while 59% of water kept for AE.

The weight of the alginate beads of AE with and without *Geobacillus stearothermophilus* reached a constant weight at 48 h (37%) and 84 h (27.4%). The result

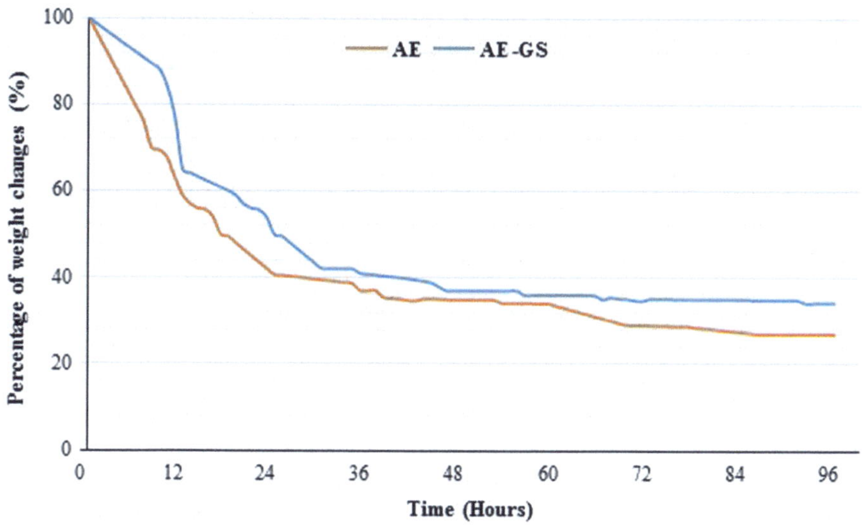


**Fig. 5.2** Weight changes on the wet hydrogel after 96 h

obtained was comparable to that study conducted by Wang et al. [13] which indicated that hydrogel retained water at 70% for first 12 h and reduced to 30% at 24 h of immersion.

It was found that the presence of bacteria in the alginate beads retained more water as compared to those alginates without the bacteria. The water retention property reflects the capability of alginate beads to keep its moisture content, therefore, maintaining water/cement ratio and at the same time acts as a water reservoir for bacterial activity when the crack occurs [13].

## 5.3 Quantification and Viability of Alginate-Encapsulated *Geobacillus stearothermophilus* in Mimicked Mortar Environment

The quantity of urease released and the viability number of bacteria, when embedded in alginate hydrogel, were assessed in the mimicked mortar environment. The results explained in the following sub-sections.

### *5.3.1 Quantification of Urease Activity in Encapsulated* Geobacillus stearothermophilus

The urease activity derived by the encapsulated *Geobacillus stearothermophilus* was measured using Total Ammonia Nitrogen (TAN) method by Nessler. Figure 5.3 shows the urea decomposition after 24 and 96 h immersed in cement slurry for both in free cell condition (without encapsulating) and after being encapsulated. The free cells recorded higher decomposition of urea at 11.9–13.0 g/L as compared to the encapsulated *Geobacillus stearothermophilus* marked 8.5–10.1 g/L at first 24 h. After immersion for 96 h, higher decomposition of urea with those of free live cells attained 20.0–23.0 g/L as compared to the encapsulated one recorded 17.0–19.0 g/L.

Table 5.2 shows a comparison between urease activity of alginate hydrogel in the present study with those investigated by previous researchers after 24 and 96 h immersed in the cement slurry. The urease activity by alginate-encapsulated *Geobacillus stearothermophilus* was lower than those reported by Wang et al. [12, 13]. However, decomposition of urease obtained at 96 h for free cells by *Geobacillus stearothermophilus* were higher than the result achieved by Wang et al. [12].

Based on the results by Wang et al. [12], both encapsulated bacteria and free cells at 96 h observed to have a same amount of urea decomposed. It shows that modified alginate hydrogel does not restrict bacterial activity and the suitability of hydrogel as the encapsulation material.

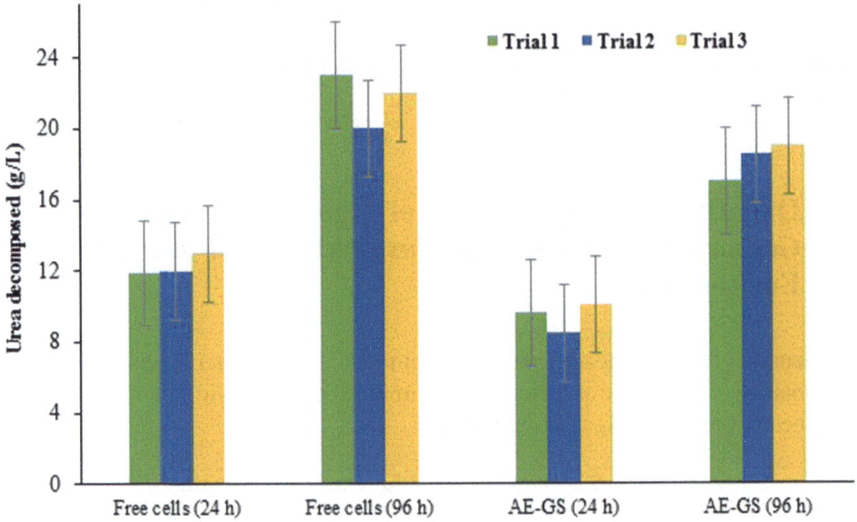


**Fig. 5.3** Urea decomposition in nutrient modified broth for first 24 and 96 h of free live cells of *Geobacillus stearothermophilus* towards the encapsulated *Geobacillus stearothermophilus* (AE-GS)

**Table 5.2** Comparison between urease activity of alginate hydrogel in the present study with those investigated by previous researchers at 24 and 96 h

| References | Type of hydrogel and bacteria | Urea decomposed (g/L) | | % Reduction in bacterial activity (urea decomposition) with respect to that of free cells. |
|---|---|---|---|---|
| | | 24 h | 96 h | |
| Raden Maizatul Aimi [8] | Free cells | 12.3 | 21.7 | |
| | Alginate hydrogel and *Geobacillus stearothermophilus* | 9.4 | 18.2 | 23.6<br>16.1* |
| Wang et al. [12] | Free cells | 17.9 | 20.0 | |
| | Modified alginate hydrogel and Bacillus sphaericus | 15.0 | 20.0 | 16.2* |
| Wang et al. [13] | Free cells | 18.0 | – | |
| | Pure hydrogel and *Bacillus sphaericus* | 15.0 | – | 16.7* |

In the other hand, results acquired from this experimental work proved that alginate-encapsulated *Geobacillus stearothermophilus* (AE-GS) still capable to produce urease even though urease activity for free cells and encapsulated bacteria at first 24 h was reduced by 23.6% compared to 16.2 and 16.7% reduction with those of other bacteria in a study conducted by Wang et al. [12, 13], respectively. Slower increment on decomposition of urea due to the fact that encapsulation network limits the bacterial activity in decomposing urease. The network of hydrogel hinder the diffusion of material to pass through [18] reflected the ability of alginate hydrogel as encapsulated material, therefore, meet the purpose on the encapsulation itself.

### *5.3.2 Viability of Encapsulated* Geobacillus stearothermophilus *in the Mimicked Mortar Environment*

The viability test was conducted to measure the cell-entrapping property by alginate-hydrogel. Using viable cell-counting method, the *Geobacillus stearothermophilus* embedded into alginate beads were countable through serial dilution method. The viability of AE-GS are illustrated in Fig. 5.4.

As shown in Fig. 5.4, the reduction of cell numbers occur between 10 and 15%. For instance, on the concentrations of $1 \times 10^{11}$ cfu/ml (T1 to T3), the reduction number occurred between $8.5 \times 10^{10}$ cfu/ml and $9.0 \times 10^{10}$ cfu/ml. The reduction

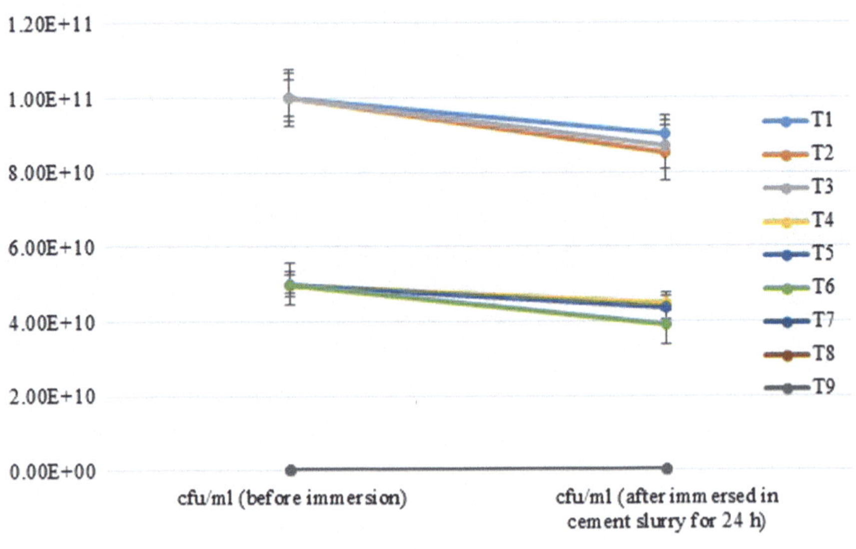


**Fig. 5.4** Viability of AE-GS containing different concentrations of bacteria cells (cfu/ml) before and after immersion in cement slurry for 24 h

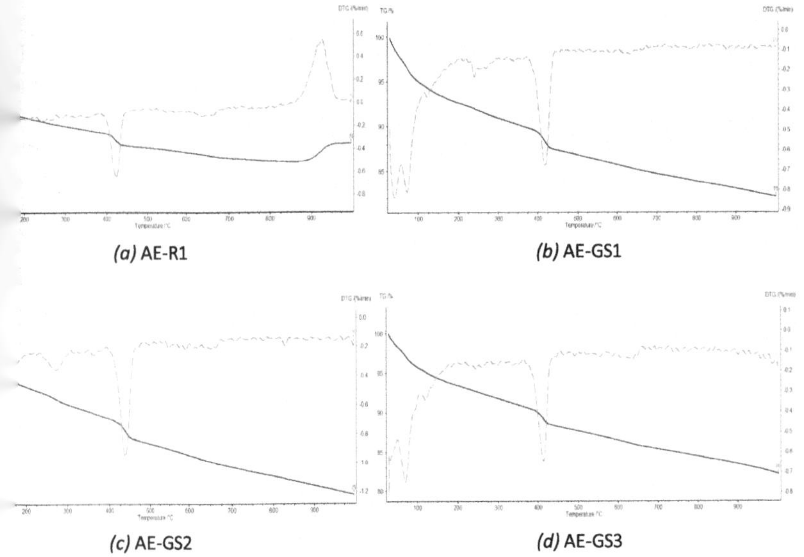


TG/DTG curves for mortar that contained alginate-encapsulated without *Geobacillus* *rmophilus* (AE-R1) and alginate-encapsulated *Geobacillus stearothermophilus* (AE-GS1– for 60 days of age

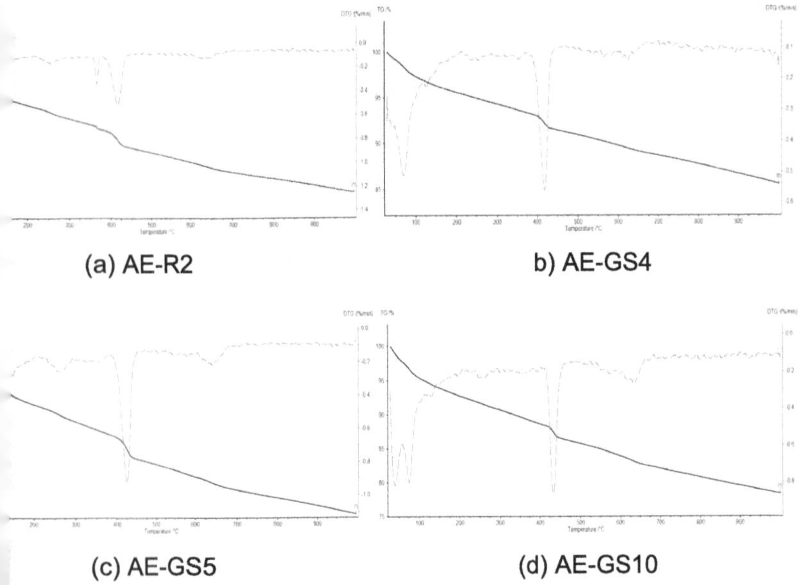


TG/DTG curves for mortar that contained alginate-encapsulated without *Geobacillus* *ermophilus* (AE-R2) and alginate-encapsulated *Geobacillus stearothermophilus* (AE-GS4–10) tested at 60 days of age

up to 4.5 × 1010 cfu/ml for the concentration of 5.0
For that concentration of 1 × $10^3$ cfu/ml (T7 to T9) th
to 9.0 × $10^2$ cfu/ml. In a study conducted by Wang e
the encapsulated spores was recorded up to 36% after 2
those reported in this study. The decrease of cells would
During the encapsulation process, the bacteria need to su
stress during stirring process [3]. The leakage of bacteria
the encapsulation work and improper coating process m
bacteria due to the network structure of the alginate hydro
to ensure cell numbers of bacteria embedded as per desi
embedded added with the addition of 15% cell number f
so those final concentrations of bacteria were achievable.

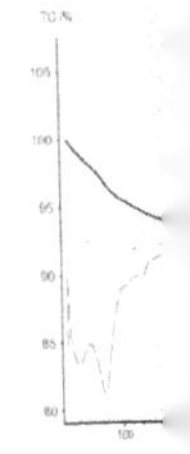

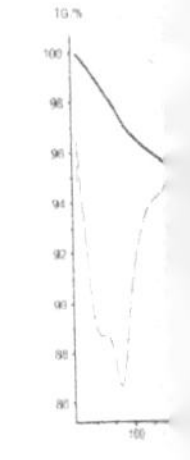

## 5.4 Thermogravimetric Analysis (TGA)

The thermogravimetric analysis (TGA) was used to observ
of mortar incorporated with alginate-encapsulated withou
*mophilus* (AE-R) and alginate-encapsulated containing
*mophilus* (AE-GS). Four (4) distinct stages of mass loss
dehydration, degradation, dehydroxylation and decomposi
The focus of this TGA test was to examine the decomposi
($CaCO_3$) which occur at the temperature interval of 650–
ysis shows two types of data; first, the percentage of weig
temperature (grey-dotted line) and second, the derivative
against the temperature (black line).

**Fig. 5.5**
*stearotl*
AE-GS

Figures 5.5, 5.6 and 5.7 shows the result for TGA of all m
age. The results from those at 60 days of age selected becau
mass of calcium carbonate ($CaCO_3$) recorded as compared
As seen in Fig. 5.5 (black line), the weight of the sample g
room temperature till about 1000 °C. The weight loss befo
dehydration of the samples. Weight loss between 200 °C a
the degradation of the alginate hydrogel.

The radius of curvature at temperature 450–500 °C (dott
thermal decomposition with the formation of gaseous rea
define the dehydroxylation process of calcium hydroxide
due to the water steaming.

Further decomposition with a small reduction of weig
of 600–750 °C. The reference specimens of AE-R1 conta
$CaCO_3$ as compared to 0.174, 0.185, and 0.196 mg of Ca
GS1, AE-GS2 and AE-GS3, respectively. The result obtai
3% replacement of AE-GS at 60 days proved that the co
concentration benefiting production of $CaCO_3$. Results p
those other two (2) levels as seen in Fig. 5.5d compared to

**Fig. 5.**
*stearo*
AE-G

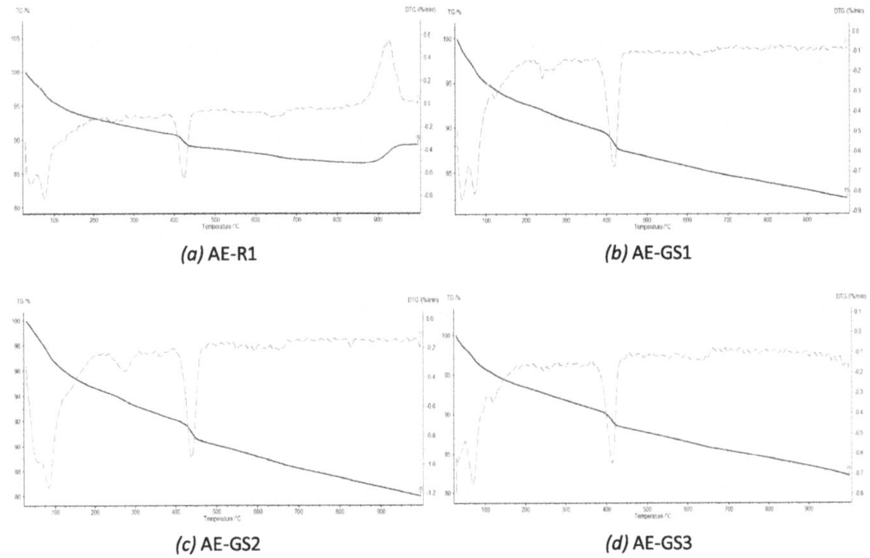


**Fig. 5.5** TG/DTG curves for mortar that contained alginate-encapsulated without *Geobacillus stearothermophilus* (AE-R1) and alginate-encapsulated *Geobacillus stearothermophilus* (AE-GS1–AE-GS3) for 60 days of age

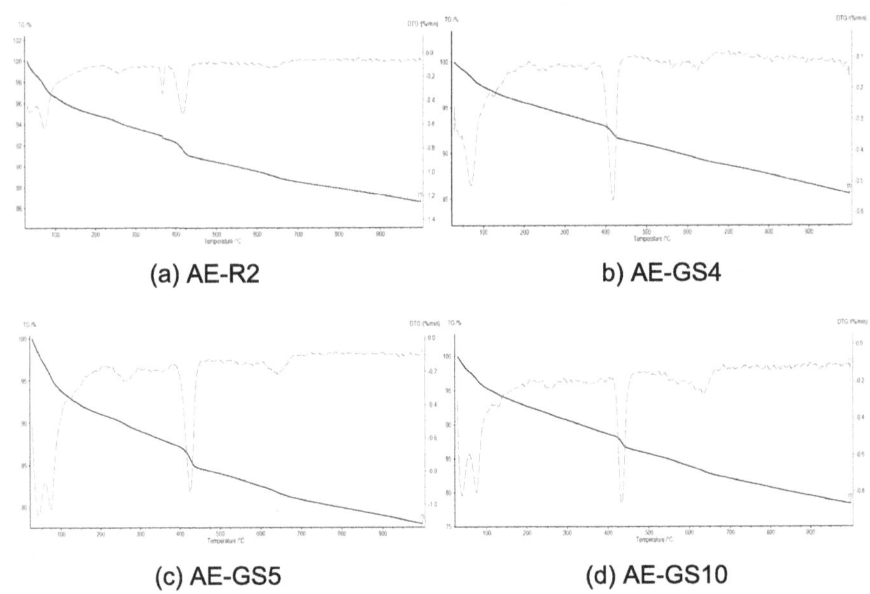


**Fig. 5.6** TG/DTG curves for mortar that contained alginate-encapsulated without *Geobacillus stearothermophilus* (AE-R2) and alginate-encapsulated *Geobacillus stearothermophilus* (AE-GS4–AE-GS10) tested at 60 days of age

up to 4.5 × 1010 cfu/ml for the concentration of 5.0 × $10^{10}$ cfu/ml (T4 to T6). For that concentration of 1 × $10^3$ cfu/ml (T7 to T9) the counted cells reduced up to 9.0 × $10^2$ cfu/ml. In a study conducted by Wang et al. [12], the reduction of the encapsulated spores was recorded up to 36% after 24 h which was higher than those reported in this study. The decrease of cells would occur due to many factors. During the encapsulation process, the bacteria need to sustain mechanical and shear stress during stirring process [3]. The leakage of bacteria would occur [3, 12] during the encapsulation work and improper coating process might release the entrapped bacteria due to the network structure of the alginate hydrogel matrix [12]. Therefore, to ensure cell numbers of bacteria embedded as per design every batch of bacteria embedded added with the addition of 15% cell number from total cfu/ml required, so those final concentrations of bacteria were achievable.

## 5.4 Thermogravimetric Analysis (TGA)

The thermogravimetric analysis (TGA) was used to observe the decomposition mass of mortar incorporated with alginate-encapsulated without *Geobacillus stearothermophilus* (AE-R) and alginate-encapsulated containing *Geobacillus stearothermophilus* (AE-GS). Four (4) distinct stages of mass loss transitions which include dehydration, degradation, dehydroxylation and decomposition (Table 4.1—Chap. 4). The focus of this TGA test was to examine the decomposition of calcium carbonate ($CaCO_3$) which occur at the temperature interval of 650–700 °C. Each TGA analysis shows two types of data; first, the percentage of weight loss plotted against the temperature (grey-dotted line) and second, the derivative of the weight loss plotted against the temperature (black line).

Figures 5.5, 5.6 and 5.7 shows the result for TGA of all mortar batches at 60 days of age. The results from those at 60 days of age selected because higher decomposition mass of calcium carbonate ($CaCO_3$) recorded as compared to 7 and 28 days of age. As seen in Fig. 5.5 (black line), the weight of the sample gradually decreased from room temperature till about 1000 °C. The weight loss before 200 °C was caused by dehydration of the samples. Weight loss between 200 °C and 450 °C was caused by the degradation of the alginate hydrogel.

The radius of curvature at temperature 450–500 °C (dotted line) also indicates the thermal decomposition with the formation of gaseous reaction product [16] which define the dehydroxylation process of calcium hydroxide that leads to weight loss due to the water steaming.

Further decomposition with a small reduction of weight occur at temperature of 600–750 °C. The reference specimens of AE-R1 contain 0.147 mg amount of $CaCO_3$ as compared to 0.174, 0.185, and 0.196 mg of $CaCO_3$ for specimens AE-GS1, AE-GS2 and AE-GS3, respectively. The result obtained from those made of 3% replacement of AE-GS at 60 days proved that the contribution of higher cell concentration benefiting production of $CaCO_3$. Results portrayed as compared to those other two (2) levels as seen in Fig. 5.5d compared to Fig. 5.5a–c.

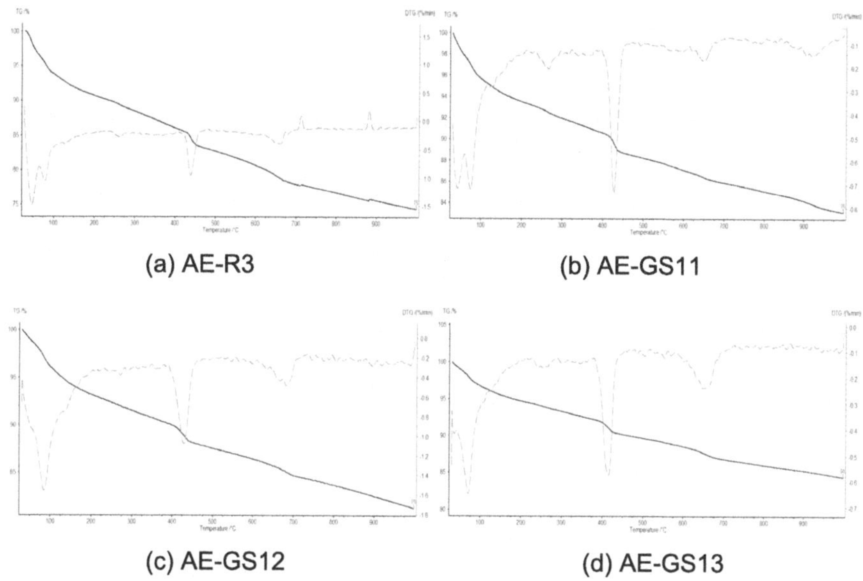


**Fig. 5.7** TG/DTG curves for mortar that contained alginate-encapsulated without *Geobacillus stearothermophilus* (AE-R3) and alginate-encapsulated *Geobacillus stearothermophilus* (AE-GS11–AE-GS13) for 60 days

Figure 5.6 shows an increased amount of $CaCO_3$ for all samples tested at 60 days of age that contained 9% of AE-GS from the total weight of mortar. Weight loss occur at temperature of 600–750 °C with the reference specimens of AE-R2 decomposition of $CaCO_3$ at 0.171 mg. Slight increase of $CaCO_3$ amount as compared to data obtained from 60 days of age for AE-GS1, AE-GS2 and AE-GS3 samples in which decomposition mass for 60 days recorded at 0.193 mg, 0.215 mg and 0.222 mg of $CaCO_3$, respectively.

It can be seen that from Fig. 5.6b–d, that at temperature of 600–750 °C (grey-dotted line), the formation of melting curve were sharper rather than that made of AE-GS at 3% (AE-GS1 to AE-GS3).

Figure 5.7 shows TGA results of all series of mortar samples containing 15% AE-GS from the total weight of mortar. The decreased in weight was observed in the sample decomposed from room temperature till about 1000 °C. Weight loss occur at temperature of 600–750 °C with the reference specimens of AE-R3. AE-R3 recorded slight increase of $CaCO_3$ by 4.7% as compared to AE-R2.

Increased of $CaCO_3$ amount as compared to data of same age for specimens of AE-GS5, AE-GS6 and AE-GS10 was observed corresponding to the decomposition mass which recorded 0.217, 0.234, and 0.245 mg of $CaCO_3$, respectively and this can be reflected by drastic reduction of weight at temperature of 600–750 °C (grey-dotted line). By comparing the specimens with and without the presence of bacteria, higher decomposition mass belong to the specimens with higher concentration of bacteria (AE-GS13, $1 \times 10^{11}$ cfu/ml).

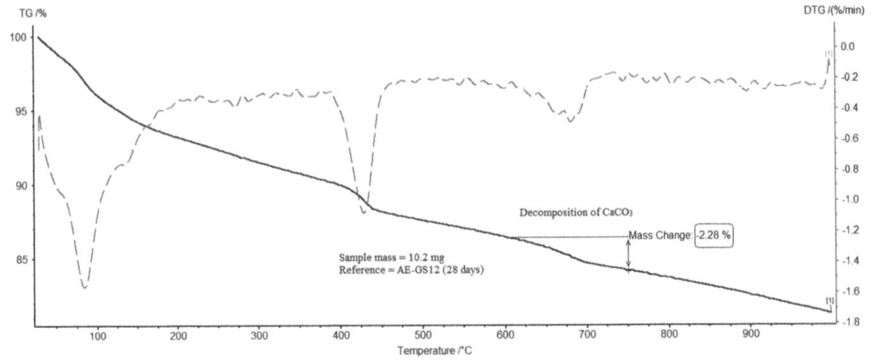


**Fig. 5.8** Decomposition mass of $CaCO_3$ at temperature of 600 °C to 750 °C. Data extracted from TGA curve, Mass of sample = 10.200 mg; decomposition percentage = −2.28%; content of $CaCO_3$ = 0.232 mg

All the decomposition mass obtained was calculated (see Fig. 5.8) and tabulated as shown in Table 5.3. By increasing the content of AE-GS in mortar mix, the $CaCO_3$ precipitation also increased. Also, a noticeable weight loss occurred on the reference specimens (AE-R1, AE-R2 and AE-R3) even though the weight loss was lower than those mortar made with AE-GS. Among those three (3) bacterial concentrations employed in this study, highest $CaCO_3$ precipitation belong to that made of $1 \times 10^{11}$ cfu/ml by a higher content of AE-GS (AE-GS13). This situation proved that a large number of bacteria cells and higher content of AE-GS are needed to produce copious amount of $CaCO_3$. The $CaCO_3$ relative decomposition mass of AE-GS mortar was compared with AE-R mortar, to measure the performance of bacteria towards precipitation of $CaCO_3$.

The results are calculated and as shown in Table 5.4. Based on the results obtained, the presence of *Geobacillus stearothermophilus* actively converting embedded nutrient in alginate hydrogel with the increase of more than 70% $CaCO_3$ content at 7 days of age.

However, at 28 days and 60 days, the decomposition occurs not as much as in 7 days. This results proved that the precipitation of $CaCO_3$ primarily at an early age and reduced by increasing of age.

The TGA in this study presented along with TGA results as reported in the literature to grasp an idea of bacteria capability in producing $CaCO_3$. At 7 days of age, investigation by Bundur et al. [2] observed 42% decomposition loss of $CaCO_3$ with the incorporation of vegetative cells *Sporocarsina pasteurii* in mortar. In another study conducted by Wang et al. [13], the decomposition mass loss of $CaCO_3$ at 26% was found by incorporation of *Bacillus sphaericus* spores in the mortar at 7 days of age. By incorporating vegetative cells at high concentration has provided more nucleation sites [15], thus conversion of nutrient became $CaCO_3$ is greater as compared with using spores. The TGA tested on mortar amended with alginate-encapsulated *Geobacillus stearothermophilus* showed that higher forms of $CaCO_3$

**Table 5.3** Decomposition mass of calcium carbonate ($CaCO_3$) content at the temperature of 650–700 °C for different age of mortar specimen containing alginate-encapsulated without (AE-R) and with different concentration of *Geobacillus stearothermophilus* (AE-GS)

| Exp. No. | Batch | Age | | |
|---|---|---|---|---|
| | | 7 days (mg) | 28 days (mg) | 60 days (mg) |
| 1 | AE-R1 | 0.076 | 0.127 | 0.147 |
| 2 | AE-GS1 | 0.132 | 0.140 | 0.174 |
| 3 | AE-GS2 | 0.139 | 0.145 | 0.185 |
| 4 | AE-GS3 | 0.143 | 0.148 | 0.196 |
| 5 | AE-R2 | 0.078 | 0.146 | 0.171 |
| 6 | AE-GS4 | 0.162 | 0.172 | 0.193 |
| 7 | AE-GS5 | 0.169 | 0.175 | 0.215 |
| 8 | AE-GS10 | 0.171 | 0.179 | 0.222 |
| 9 | AE-R3 | 0.094 | 0.154 | 0.179 |
| 10 | AE-GS11 | 0.184 | 0.210 | 0.217 |
| 11 | AE-GS12 | 0.231 | 0.219 | 0.234 |
| 12 | AE-GS13 | 0.220 | 0.232 | 0.245 |

**Table 5.4** Calcium carbonate ($CaCO_3$) decomposition mass in percentage (%) between AE-GS mortar with AE-R (control)

| Exp. No. | Batch | Age | Exp. No. | Batch |
|---|---|---|---|---|
| | | 7 days (%) | 28 days (%) | 60 days (%) |
| 1 | AE-R1 | – | – | – |
| 2 | AE-GS1 | 73.7 | 10.2 | 18.4 |
| 3 | AE-GS2 | 82.9 | 14.2 | 25,9 |
| 4 | AE-GS3 | 88.2 | 16.5 | 33.3 |
| 5 | AE-R2 | – | – | – |
| 6 | AE-GS4 | 107.7 | 17.8 | 12.9 |
| 7 | AE-GS5 | 116/7 | 19.9 | 25.7 |
| 8 | AE-GS10 | 119.2 | 22.6 | 29.8 |
| 9 | AE-R3 | – | – | – |
| 10 | AE-GS11 | 95.7 | 36.4 | 21.2 |
| 11 | AE-GS12 | 126.6 | 42.2 | 30.7 |
| 12 | AE-GS13 | 134.0 | 50.6 | 36.9 |

was observed compared to those found from the studies above, reflected by more top % of decomposition mass loss (42–57%) of $CaCO_3$.

The higher decomposition mass of $CaCO_3$ obtained when 15% AE-GS from the total weight of mortar was incorporated, and the same dosage is expected required for maximum healing efficiency.

In sum, swelling capacity and retention properties on alginate-encapsulated *Geobacillus stearothermophilus* hydrogel suggested that the alginate hydrogel is suitable to be used as a bacterial carrier. It also discovered that *Geobacillus stearothermophilus* was compatible to be encapsulated in alginate hydrogels. Based on the swelling test and water retention properties, the alginate-encapsulated with *Geobacillus stearothermophilus* have a minimum swelling ratio thus created a minimal void that will not disturb the mortar characteristic. The alginate-encapsulated *Geobacillus stearothermophilus* also able to retain water longer as compared to alginate-encapsulated without *Geobacillus stearothermophilus* inferring that gels able to remain intact before it breaks. Apart of that, the quantification of urease activity and the availability of *Geobacillus stearothermophilus* after being encapsulated result obtained were comparable to the result reported in the literature which highlighted the suitability of this bacteria used in self-healing concrete.

The quantification of urease activity was performed using free live cells of *Geobacillus stearothermophilus* with the encapsulated one. Higher concentration of urea measured with the free cells obtained 20.0–23.0 g/L as compared to the encapsulated *Geobacillus stearothermophilus* at 17.0–19.0 g/L. The lower result of urease decomposition in encapsulated *Geobacillus stearothermophilus* was because alginate hydrogel networking hinders any substances to pass through, thus, reflected the ability of alginate hydrogel as encapsulated material. This result also indicated the capability of *Geobacillus stearothermophilus* to produce urease when embedded into mortar since *Geobacillus stearothermophilus* is facultative anaerobic types of bacteria (bacteria that can grow with or without the presence of oxygen). On the other hand, the reduction of cell numbers was up to 15% on the viability test. Therefore, to ensure the initial concentration used for encapsulation process, for every batch of encapsulated bacteria, 15% of an additional cell number would be added to make up its final concentration as per design.

The results obtained from TGA analysis showed that the higher concentration of bacteria with the higher replacement of AE-GS in mortar has contributed to a higher dosage of calcium carbonate precipitation. The presence of bacteria showed distinctive decomposition rate of $CaCO_3$ due to high temperature as compared to con- trol specimens without the bacteria.

## References

1. Albasri, H.M., Almohammadi, A.A., Alhhazmi, A., Bukhari, D.A., Waznah, M.S., Mawad, A.M.: Production and characterization of rhamnolipid biosurfactant from thermophilic *Geobacillus stearothermophilus* bacterium isolated from Uhud Mountain. Front. Microbiol. **15**, 1358175 (2024)
2. Bundur, Z.B., Kirisits, M.J., Ferron, R.D.: Biomineralized cement-based materials: impact of inoculating vegetative bacterial cells on hydration and strength. Cem. Concr. Res. **67**, 237–245 (2015). https://doi.org/10.1016/j.cemconres.2014.10.002

3. Callone, E., Campostrini, R., Carturan, G., Cavazza, A., Guzzon, R.: Immobilization of yeast and bacteria cells in alginate microbeads coated with silica membranes: procedures, physico-chemical features and bioactivity. J. Mater. Chem. **18**(40), 4839–4848 (2008)
4. Cook, M.T., Tzortzis, G., Charalampopoulos, D., Khutoryanskiy, V.V.: Production and evaluation of dry alginate-chitosan microcapsules as an enteric delivery vehicle for probiotic bacteria. Biomacromolecules. **12**(7), 2834–2840 (2011)
5. Georget, E., Kapoor, S., Winter, R., Reineke, K., Song, Y., Callanan, M., Ananta, E., Heinz, V., Mathys, A.: In situ investigation of *Geobacillus stearothermophilus* spore germination and inactivation mechanisms under moderate high pressure. Food Microbiol. **41**, 8–18 (2014)
6. Lee, K.Y., Mooney, D.J.: Alginate: properties and biomedical applications. Prog. Polym. Sci. **37**(1), 106–126 (2012). https://doi.org/10.1016/j.progpolymsci.2011.06.003
7. McMullan, G., Christie, J., Rahman, T., Banat, I., Ternan, N., Marchant, R.: Habitat, applications and genomics of the aerobic, thermophilic genus *Geobacillus*. Biochem. Soc. Transit. **32**(2) (2004)
8. Raden Maizatul Aimi, M.A.: Enhancement of Autonomous Healing Mortar through Encapsulated Bacteria Biomineralisation. Ph.D. thesis,. Universiti Teknologi MARA, Shah Alam, Selangor, Malaysia (2016)
9. Rogers, J.V., Sabourin, C.L.K., Choi, Y.W., Richter, W.R., Rudnicki, D.C., Riggs, K.B., Taylor, M.L., Chang, J.: Decontamination assessment of bacillus anthracis, bacillus subtilis, and *Geobacillus stearothermophilus* spores on indoor surfaces using a hydrogen peroxide gas generator. J. Appl. Microbiol. **99**(4), 739–748 (2005). https://doi.org/10.1111/j.1365-2672.2005.02686.x
10. Stradiotto, G.C., Heckler, C., Chincha, A.A., Sant'Ana, A.S.: The fate of spore-forming bacteria throughout the gelatin processing. Int. J. Food Microbiol. **111355** (2025)
11. Valenzuela, B., Solís-Cornejo, F., Araya, R., Zamorano, P.: Isolation of thermophilic bacteria from extreme environments in northern Chile. Microorganisms. **12**(3), 473 (2024) https://www.mdpi.com/2076-2607/12/3/473
12. Wang, J., Mignon, A., Snoeck, D., Wiktor, V., Van Vliergerghe, S., Boon, N., De Belie, N.: Application of modified-alginate encapsulated carbonate producing bacteria in concrete: a promising strategy for crack self-healing. Front. Microbiol. **6**(1088), 1–14 (2015). https://doi.org/10.3389/fmicb.2015.01088
13. Wang, J., Snoeck, D., Van Vlierberghe, S., Verstraete, W., De Belie, N.: Application of hydrogel encapsulated carbonate precipitating bacteria for approaching a realistic self-healing in concrete. Constr. Build. Mater. **68**, 110–119 (2014)
14. Watanabe, T., Furukawa, S., Hirata, J., Koyama, T., Ogihara, H., Yamasaki, M.: Inactivation of *Geobacillus stearothermophilus* spores by high-pressure carbon dioxide treatment. Appl. Environ. Microbiol. **69**(12), 7124–7129 (2003)
15. Wei, S., Cui, H., Jiang, Z., Liu, H., He, H., Fang, N.: Biomineralization processes of calcite induced by bacteria isolated from marine sediments. Braz. J. Microbiol. **46**(2), 455–464 (2015)
16. Widmann, G.: Interpreting TGA curves, vol. 1, p. 20. Mettler Toledo GmbH, Analystical, Sonnenbergstrasse 74, CH-8603 Schwerzenbach, Switzerland (2001). http://www.masontechnology.ie/x/Usercom_13.pdf
17. Xue, L., Yang, N., Xu, X., Jin, Y., Cao, X., Zhang, H.: Inactivation of *Geobacillus stearothermophilus* spores by induced electric field in different food mediums. Food Microbiol. **104832** (2025)
18. Yang, T.: Mechanical and swelling properties of hydrogels. Fibre and Polymer Technology, Coating Technology (2012)

# Index





R. M. A. M. Azam and D. H. M. Saman, *Self-Healing Concrete*, Engineering Materials,
https://doi.org/10.1007/978-981-95-7553-4

The manufacturer's authorised representative in the EU is Springer Nature Customer Service Centre GmbH, Europaplatz 3, 69115 Heidelberg, Germany. If you have any concerns regarding our products, please contact ProductSafety@springernature.com

Printed and bound by CPI Group (UK) Ltd, Croydon, CR0 4YY
21/09/2026
02219968-0006